RATIONALLY TURBULENT EXPECTATIONS

CAPITAL MARKETS REINTERPRETED

KENT OSBAND

Rationally Turbulent Expectations: Capital Markets Reinterpreted

For information about this title or to order other books and/or electronic media, contact the publisher:

Kent Osband
Mountain Brook, AL
www.risktick.com
info@risktick.com

Hardcover ISBN: 978-1-7343376-4-8
Paperback ISBN: 978-1-7343376-5-5
Electronic ISBN: 978-1-7343376-6-2

Printed in the United States of America
Cover graphic by Katrin Primak, Dreamstime.com
Cover design by Woven Red Author Services

First published 24 March 2025
Last updated 19 November 2025

Contents

Preface

This book rests on a simple observation: predicting future risks and rewards is hard. We can't deduce their evolution from neat physical laws. We can't sample observations from the future.

Future returns on capital are particularly opaque. So much depends on products that haven't yet been invented, rivals that haven't yet emerged, regulations that haven't yet been imposed, and disasters that haven't yet struck. Although the capital market marshals a lot of brainy effort, it never sees clearly what lies in store. Instead it observes the profits and perils it weaved through before and guesstimates what might come next. In short, it looks backward into the future.

Backward into the future? In most languages that sounds bizarre. Words like "ahead" and "forward" that point to space in front of us also point to future time. However, that isn't the only way to align directions. A handful of languages link by what we observe. The most prominent is the Incan language Quechua. In Quechua, *ñawpa* means both "ahead" in space and "past" in time while *qhipa* means both "behind" and "future".

Looking backward into the future is useful since the future always recapitulates some of the past and usually recapitulates most. Most surprises are outliers from a still intact trend. We tweak the next round of forecasts and move on. Occasionally the surprises get under our skin. They shock us less by their size than their persistence. They make us suspect that what we thought of as a rare outlier is now the new norm.

Imagine you're driving a car backward and all you see clearly is the road already traveled. Wouldn't you look more closely at recent twists and turns, to gauge whether you're heading off course? Wouldn't you watch other cars more carefully and tend to mimic their moves? Wouldn't you be more tentative and make frequent corrections, many of which you quickly reverse? Noisy learning would make your forecasts volatilely volatile: sometimes calm, sometimes stormy. Call it RATE for Rationally Turbulent Expectations. In contrast, the Full-Information Rational Expectations (FIRE) paradigm that dominates macroeconomics treats the future as a well-lit freeway.

The core differences can be summarized in a few equations, which I will try to make clear and intuitively appealing. We will then apply those equations to neatly solve a host of long-standing finance puzzles. Perhaps not solve them perfectly, but well enough to persuade that the RATE paradigm handily beats FIRE.

Will this help you make money trading? Not without more detailed knowledge and financial engineering skills than I offer. This book examines why social learning behaves the way it does, not how one might outsmart it. However, I hope it conveys some useful insights, including:

- Investors aren't as schizoid as the prevailing memes suggest.
- Doubts and disagreement speed learning.
- Big disagreement is better treated as a call for deeper investigation than as wrong think to be squelched.
- Fast learning about new trends is bound to be turbulent.
- Capital markets are poor long-term prophets but great short-term error correctors.
- Regulators should focus less on dampening short-term volatility and more on assuring long-term sustainability.
- Big duration mismatch in banks is inherently destabilizing.
- Massive sovereign debts pose far graver risks than bond markets currently suggest.

Abbreviations

AI	artificial intelligence
ACF	autocorrelation function
APL	Almost-Perfect Learner
	bond stock or share of GDP B,b
BR	Beta Reset
bps	basis points
c	consumption
c^*,C	credit spreads
CDS	credit default swaps
CRRA	constant relative risk aversion
D_{eff}	effective defaults or gamma shape
E	mean belief
EMA	exponential moving average
EVSI	expected value of sample information
F	market E
f	Kelly fraction
GARCH	Generalized Auto-regressive Conditional Heteroskedasticity
GDP	gross domestic product
g	GDP growth rate
h	objective probability density or hazard rate
h^*	stationary probability density
HHI	Herfindahl-Hirschman Index
iid	independent, identically distributed
ITM	inter-threaded market
J	pure dividend yield
K,k	capital / share of total capital
log	logarithm / logarithmic
N	pool size
NPV	net present value
p	subjective probability
P	equity price or ratings transition matrix
pct	percent
PL	Perfect Learner
Q,q	price-to-dividend ratio / multiplier

q	primary surplus / B
R,r	pct/log discount rate
r_p	equity risk premium
RIL	relative information loss
RMSE	root mean squared error
S	Sharpe ratio or survival rate
SV	stochastic volatility
s	diffusion speed
S&P	Standard & Poor's
TD	Tiny Doubters
T,t	time
T_{eff}	effective sample size
U	information measure
US	United States
V	variance of beliefs
vol	volatility
x	observed outcome
y	standardized news
z	Brownian noise or evidence on risk

Greek Characters

α,β	beta distribution parameters
α	sensitivity of default rate to bond stock
γ	relative risk aversion
Δ	change
δ	$r-g-q$
η	learning rate or pct loss on default
θ	Bernoulli risk
I	identity matrix
K	cumulant generating function
κ	cumulant
Λ,λ	actual or perceived reset rates
M	moment generating function
μ	drift of Brownian x
ν	aggregate default risk
ξ	consumption/savings tradeoff multiplier
Π	switching rate matrix
π	reset distribution or Markov switching rate
ρ	rate of time preference
σ	volatility of x
τ	scaling of volatility
ς	equity price volatility
ψ	intemporal elasticity of substitution
Ω	Debt Apocalypse
ω	credit risk multiplier

0

Overview

Contemporary finance theory rests on an unstable truce between two opposing schools. While both schools agree that the capital market discounts streams of inherently risky returns, they disagree on how it functions. The Rational Expectations school treats the market as a knowledge machine that assesses risks correctly and prices them appropriately. The Behavioral Finance school treats the market as a ship of fools prone to long stretches of complacency and short bouts of panic.

The truce categorizes each financial behavior by whichever theory describes it best, so in some ways it cannot be refuted. Nobel Prizes have been awarded on each side. But there is one core question the truce cannot answer. Given market competition, why don't knowledgeable investors eventually squeeze the rest out?

The two schools err in opposite directions. Standard Rational Expectations is super-rational. Its agents forecast most future trends well even though the future cannot be directly sampled and might differ greatly from past trends. Noisy data never misleads, changes in trends never confuse, and the true odds of any bet are never concealed. Capital markets convincingly refute this. But Behavioral Finance is far too quick to ascribe the violations to "irrational exuberance" and other foolish emotional drivers. Market traders aren't gambling in brightly lit casinos; they're playing blackjack in the dark. Their games are bound to trigger misplays and changes in strategy.

This book presents a unifying paradigm. It treats the capital market as a rational learning machine. Rational rather than foolish because the market incentivizes good short-term predictions and quickly corrects recognized errors. Learning rather than knowing because risks are inherently too unstable to reliably forecast long-term trends.

The most striking difference between learning and knowing is how expectations behave. Super-rational expectations, sometimes described as Full-Information Rational Expectations or FIRE, are relatively stable. In contrast, learning about unstable risks induces continual, noisy corrections. Fast learning can make expectations far more volatile than the trends they aim to predict. This volatile volatility, akin to turbulence in fluids and accounts for a host of seemingly foolish behavior. Accordingly, I call the unifying paradigm Rationally Turbulent Expectations or RATE.

RATE models beliefs as subjective probability distributions p over various hypotheses about the world, which people test against the flow of observations x. The core learning equation has two distinctive parts. The first is standard Bayesian updating, which revises p in proportion to hypotheses' likelihoods of generating the latest x. The second is doubt about the stability of risk, framed as tiny migrations of p across hypotheses. Most notably, the mean belief E adjusts faster to news when variance V is higher. as high V downweights previous conviction.

FIRE misses this because it presumes that trends are stable enough to drop the second part of the core learning equation. Absent migration, V tends to shrink inversely to the number of observations. Eventually V gets small enough to treat p as certain. While FIRE is not hostile to learning, it assumes that learning was so effective that there is little scope for more.

In real-life markets, both V and fluctuations in V have significant impacts. High V induces extra emphasis on recent evidence. This reflects reasonable doubts about the relevance of older observations and the inability to sample the future. It makes rational learners look myopic and makes prices more volatile. Fluctuations in V induce variable responses as the value of information depends on context. They make rational learners look complacent at some times and panicky at others.

Mathematically, the relevant summary features of p are cumulants. Cumulants are higher moments stripped of most influence from lower moments. The Core Learning equation indicates that the volatility of each cumulant is driven by the cumulant one order higher. The resulting hierarchy induces bifurcated behavior, where disturbances either dissipate over time or resonate chaotically.

Cumulant hierarchies occur in many complex physical systems. The most dramatic examples involve fluids, which flow smoothly at low velocities and turbulently at high velocities. Financial markets too are often described as turbulent (a Google search for "financial turbulence" returns 43 million results), and the description seems apt apart from doubts about financial vortices.[1]

Turbulence has an unnerving reputation. It flares unexpectedly, changes quickly, bodes danger, and is hard to predict. Yet it stems from something useful: rapid flow for fluids, rapid learning for beliefs. The challenge is to manage it appropriately. Combustion engines turbulently mix fuel with air to speed ignition. Golf balls are dimpled in ways that magnify turbulence but push it behind; this helps them fly straighter and faster. Rational turbulence is best managed by respecting the divergence of opinion and investigating its causes with open minds. New information is most valuable when turbulence is high.

RATE allows for complex dynamics that FIRE cannot fathom. But the Core Learning equation by itself can explain only individual behavior. To extend this to markets, we need another equation, which I call the Invisible Mind theorem. It demonstrates that decentralized capital markets can approximate optimally centralized learning, even though they transfer far less information. The key element is a gambling strategy known as the Kelly criterion.

Combining the two equations lets us answer a host of puzzles about financial markets. Why do they trade so much? Why are they so volatilely volatile? Why do they predict long-term trends so poorly? Why do they recover so quickly after disasters? Why do bear markets tend to be more volatile than bull markets? Why are risk premia so high? Why are bond markets prone to crises. Why do sovereign lenders seem fickle?

Outline

While my treatment is mathematical, it should be easy to follow for anyone familiar with modern statistics and finance theory. Markov processes, Brownian motion, and basic stochastic calculus are as hard as it gets. The biggest challenge is getting comfortable with the frequent layering of probabilities: subjective beliefs on top of objective risks and sometimes beliefs about beliefs. FIRE tends to collapse them into a single layer, on the grounds that rational investors eventually figure out what the risks really are. But it is only justified when underlying trends are so stable for so long that doubts don't matter. Real-world finance is full of doubt.

Chapter 1 addresses the quantification of doubt. This has long been an academic battlefield, with some claiming doubt is too murky to quantify and others claiming that a single bet can reveal all. The truth is more nuanced. Suppose we frame a series of risky events as random draws with unknown probability θ. While a full specification of our beliefs $p(\theta)$ might not be feasible, the bets we place on those events reveal key features of p. Our bet on a single outcome x indicates the mean E. Our bets on two independent, identically distributed (iid) outcomes identify the variance V in addition to E. Our bets on N iid outcomes identify the first N moments/cumulants of p.

Suppose we stagger the bets. As explained in Chapter 2, the change ΔE in the second bet tends to equal V times the unexpected news $x_1 - E$ after normalization by the objective variance σ_E^2. If ΔE is always zero, then $V = 0$ and we must be certain about p. That is the standard FIRE assumption. FIRE is wholly appropriate when p is formulated from first principles, say equal chance of every feasible outcome. FIRE is generally appropriate when we have many observations on a relatively stable distribution. However, in most economic forecasting V is too big to ignore, both because there are too few clearly relevant x to identify the current θ and because the future θ might change. The higher V is, the quicker we adjust. To put it another way, information is most valuable when V is high as we are prepared to learn more from it.

Chapter 3 shows that even tiny doubts about stability can have huge impact. Suppose that after 10,000 apparently fair coin flips, we start observing nothing but heads. At what point do we suspect something is seriously amiss? If we are certain the coin is fair, nothing changes. Standard Bayesian updating won't help much either. Since it assumes each flip is iid, even 50 straight heads wouldn't shift E by more than 0.0025. Either response looks dumb or crazy, since the chance of a fair toss generating 50 straight heads is one in a quadrillion.

The betting path looks far more reasonable if we allow some tiny chance λ, say one in a million, that the coin at one round gets switched at the next. While the optimal response varies with the imagined details, it nearly always exhibits three distinct phases: an extended period where nothing seems to change, followed by a spurt of intense change with very high V, followed by a long taper as E approaches 1. Despite our rational updates, an outside observer would likely infer that we initially ignored the evidence and then overreacted in panic.

Might this be an artefact of modeling only a single extreme shock? To check, I reformulate the betting game as a repeated Markov process and track fair prices for "Perfect Learners", who know the driving parameters and evaluate all information efficiently. Prices often look as if the market has lost its bearings: too jittery overall and veering between complacency and panic. Yet there is nothing irrational about Perfect Learners. They down-weight older information because it might have lost relevance and rightly value new information more in some contexts than in others.

Chapter 4 turns to equity markets, with dividends modeled as geometric Brownian motion with randomly changing drift and relative volatility σ. Its Core Learning equation embeds an infinite chain that can be expressed very neatly but has chaotic implications. The volatility of every cumulant κ_n of order n equals $|\kappa_{n+1}|/\sigma$. When Perfect Learners price the expected streams of dividends, two striking results immediately follow. One is excess volatility of equity prices relative to dividends. The other is fluctuating volatility, better known as GARCH behavior. These two phenomena should always be connected, so if one is ever observed without the other it will be strong evidence against the theory. The cumulant hierarchy also implies that volatility is stochastic, unlike GARCH which presumes that volatility responds deterministically to past noise.

Excess volatility is mainly driven by V, which tends to be higher in rare states than in common states. In modern economies, expansion is the norm. Hence RATE predicts that bear markets will generally be more volatile than bull markets and that volatility will tend to rise as prices fall. This is well confirmed in practice, although often attributed to fears of bankruptcy. RATE doesn't need this crutch any more than it needs irrational exuberance. RATE also accounts well for the observed long memory of volatility.

RATE doesn't insist that rational learners agree on fair price; it allows them to bet against each other. Chapter 5 investigates this. Gamblers seeking to maximize long-term wealth should eventually dominate rivals. This favors the Kelly criterion, which allocates gambling capital in proportion to the ratio of expected profit to variance. When all investors apply this strategy, the market behaves like a single rational learner who assigns probabilities in proportion to capital. Small side investments, taxes with redistribution, or shuffling of beliefs provide analogues to tiny doubts. While most gamblers temper their risks by betting a fraction f of the full Kelly bet, fractional Kelly markets behave similarly to full Kelly markets and can facilitate learning from others.

Decentralized learning via markets can save enormously on costs of centralized information collection and processing. Since human brains are not wired to single control centers, perhaps they too apply market-like mechanisms to reach joint decisions. Market-like mechanisms might also save on the mounting energy costs of AI.

The most perplexing feature of real-life-capital markets is the huge volume of trading they generate. FIRE cannot explain this as its agents think too much alike, while FIRE/behaviorist mixtures strain to generate sustainable equilibria. Chapter 6 resolves this conundrum. Simulations of RATE naturally generate a lot of trading. Their key predictions about trading patterns are well supported empirically, including the significant correlation between trading volume and volatility and the long memory of trading volume.

Heterogeneity persists mainly because capital markets are far more efficient at correcting short-term forecast errors than at gauging long-term structural views. The latter are tested less against long-term outcomes than against the changes in long-term views. This recalls Keynes's characterization of the capital market as a beauty contest that votes not on greatest beauty but on who the judges will select. Momentum trading likely seeks to profit from trends in structural views that aren't fully captured by current prices.

Chapter 7 explores the long-standing puzzle over equity risk premia. Standard models treat aggregate equities as claims on aggregate consumption that investors with constant relative risk aversion (CRRA) demand a premium to hold. When consumption growth is assumed to be iid Gaussian, the risk aversion needed to justify the observed premium is off the charts in cowardice. Incorporating occasional disaster risks in consumption roughly halves the implied CRRA but still leaves it implausibly high. Furthermore, equities are predicted to behave more like safe-haven investments than optimistic bets on growth, unless we assume intertemporal preferences that empirical evidence strongly rejects.

In reality, corporate profits comprise less than a quarter of GDP and the dominant investors have little interest in consuming them; their primary focus is long-term enrichment. Equities are far more realistically modeled as claims on dividends sought by fractional Kelly gamblers. This resolves the core puzzle. Risk premia vary endogenously, which induces extra volatility that boosts the average premium. The lower the Kelly fraction, the higher the average risk premium it can account for. Simulations suggest that the applicable Kelly fraction exceeds one-half, which is only modestly risk averse. In large developed equity markets, we should expect risk premia and equity market variance to have similar orders of magnitude, and that is what the empirical evidence shows.

Chapter 8 turns to credit markets, with defaults modeled as Poisson events. Rational learning about default risk can be interpreted as a counting game in which D_{eff} effective defaults are recorded in T_{eff} effective time. Doubts tend to diminish both D_{eff} and T_{eff}, which breeds extra volatility. Spread changes should generally skew upward and this is confirmed in market data. When mean-reverting diffusions in average default risk ν are fitted to evidence on average corporate default rates and pool sizes N, D_{eff} for Perfect Learners averages 5 while duration T_{eff}/N averages less than 4 months. The predicted spread volatilities of our benchmark model broadly conform with empirical evidence.

Chapter 9 examines credit grades, which can be viewed as multipliers ω applied to ν. Empirically, each full-letter drop in credit grade roughly quadruples ω. Defaults for top grades are so rare that we must project their ω from default rates in lower grades. These projections explain the very high correlations across credit spreads.

Credit grades tend to deteriorate over time. Expected ω rises substantially on a 10-year horizon for all but the lowest rating tiers. Top-rated credits generally have only a few percent chance of default on a 15-year horizon. However, high-grade defaults tend to be highly correlated. The threat of occasional big mark-to-market losses on high-grade portfolios might explain their seemingly high relative risk premia.

Clustered default risk is particularly salient for highly leveraged banks. Their big duration mismatch—borrowing short to lend long—is far more dangerous than most regulators realize. It can generate financial bubbles big enough to trigger a broad crisis when they bust, as mortgages did in 2008.

Chapter 10 examines sovereign risks. Reinhart and Rogoff (2009) note that sovereign debt markets seem to behave extremely myopically. Most of this can be explained by the relatively small pool of sovereigns. The mean jump on default for the sovereign benchmark is eight times the mean jump for the corporate benchmark. Only 4% of sovereign D_{eff} exceed 4, while 70% are less than 2 and over a third are less than 1.

Market confidence in sovereign servicing is partly self-fulfilling, since it allows sovereigns to roll over debt at low cost. This partly decouples default risk from economic fundamentals and allows debt bubbles to grow for decades. However, the prospects for long-term servicing ultimately hinge on sovereigns keeping their debt burden under control. Policymakers need to be extra vigilant as debt markets are likely to give little advance warning of crisis. US federal debt exemplifies the tensions.

1

Measures of Doubt

The truths we know best transcend the experiences that helped reveal them. Consider Euler's Identity $e^{\pi i}+1=0$, often called the most beautiful equation in mathematics. The 1 refers to a unit quantity detached from any other quality, a concept few species comprehend. The notion of 0 as a number developed from its use as a placeholder and is less than 2000 years old. The notion of two quantities adding to zero required negative numbers, which mathematicians long abhorred. The i was even odder as it squared to -1 and could not fit on a standard number line; only later was it reconceived as a multiplier that rotated 90 degrees without changing length. The e arose from analysis of constant proportional growth.

Euler's Identity has been known for less than 300 years. To this day most people would find it gobbledygook. It is impossible to prove empirically as neither e nor π can be precisely computed. Nevertheless, Euler's Identity is true. Absolutely, necessarily true given the axioms. No observation can refute it. No doubts are warranted.

Doubts enter when we assert that some mathematical axiom applies to the real world and use it to make predictions. When sufficiently well-tested, it is considered a scientific law. Yet well-tested can never rule out new ways to refute theory or tiny deviations that point to amendments.

Physics has learned to cherish the tiny doubts that emerge when observations defy theoretical predictions or two predictions clash.

Einstein's theory of special relativity emerged from trying to reconcile the constant speed of light, as predicted by Maxwell's equations of electromagnetism and confirmed in experiments, with Newtonian laws of relative motion. Einstein's theory of general relativity reformulated gravity as curvature of space-time. While the deviations from Newton's Laws are insignificant for most mechanics on Earth, relativistic corrections are crucial to making satellite navigation systems work.

Even in physics, it is hard to make reliable predictions when too many factors enter in with too complex interactions. Despite great advances in meteorology thanks to better modeling with supercomputers, it is hard to predict the intensity and course of hurricanes. These difficulties are multiplied many-fold in forecasts of human affairs.

Some doubts reflect a real or perceived indeterminacy of outcomes. This indeterminacy is typically known as chance or randomness in mathematics, as objective risk in economics, or as aleatory uncertainty in machine learning. Other doubts reflect gaps in knowledge, known variously as ignorance, missing information, subjective uncertainty, or epistemic uncertainty. Out of old habit I favor the objective/subjective labels, although I realize that objectivity is often just a subjective presumption. All I ask of readers is to acknowledge what Ülkümen et al (2016) called the "two dimensions" of uncertainty and to work with me to try to quantify them.

The single best indicator of people's uncertainty about an outcome x is their mean or expected belief E. There are two basic ways to elicit it. Consider a gamble that pays out $x=1$ if a specified event occurs and $x=0$ if it doesn't. Then the maximum someone would voluntarily pay for x is a good proxy for E. A crisper method rewards $S(E,x)$ (with penalties when S is negative) based on reported E and the subsequently observed x, where the expected value of S is always maximized by reporting E truthfully. Such S are called proper scoring rules, or strictly proper if deceit never ties for best (Gneiting and Rafferty 2007). An example is $S(E,x)=2Ex-E^2$. Risk aversion can be addressed through random payoffs (Karni 2009).

E does not suffice to measure doubt. In a famous example from Ellsberg (1961), suppose we draw blindly from an urn of 100 balls, identical except for color, with $x=1$ if we pick one of the N red balls and 0 otherwise. If certain that $N=50$, we estimate $E=\frac{1}{2}$ with confidence that this is the true probability θ of red. If we consider any feasible N equally likely, our beliefs about θ are highly dispersed but we still estimate $E=\frac{1}{2}$.

When betting on any single outcome, the mean belief E should predominate regardless of dispersion; we might as well imagine that $\theta=E$. However, dispersion has a huge impact on how E evolves over time. Suppose we return the chosen ball to the urn and draw blindly again. If we are certain about N, we will estimate $E=N/100$ at both rounds regardless of the outcome. If we think that N is either 0 or 100 with no chance of falling between, then after observing the first x our E will shift to match it. Higher dispersion always makes mean beliefs more sensitive to news.

Is high dispersion good or bad? That depends on context. When the underlying risk is stable, Bayesian inference will eventually squeeze convictions toward the true θ and we will wish that beliefs had narrowed sooner. In dynamic environments the relevant risk often changes, putting narrow beliefs at a disadvantage. Better to entertain diverse hypotheses than to insist on one.

This book explores the dispersion of beliefs and how it evolves as people learn. The hallmark feature is a volatile volatility that is analogous in some respects to physical turbulence. Like physical turbulence, rational turbulence often frightens the people who experience it and occasionally causes significant damage. Yet fundamentally it is just a sign and facilitator of rapid learning. Societies that respect rational turbulence flourish far better than those that try to suppress it.

For physical analogy, think of beliefs as mental fluids that broadly conform to the history that contains them but are frequently buffeted by doubts. Learning is the flow of beliefs as history reshapes. When history evolves slowly or seems to stand still, learning is calm. When a flurry of evidence makes it hard to distinguish major new trends from random outliers, beliefs get jostled so much that they tumble over each other and stir turbulence.

Terminology

I treat uncertainty as a subjective layer of beliefs on top of a layer of unknown objective risk, with each layer described by probabilities. Here is a brief elaboration along with some related terms:

[Objective] Risk: the indeterminacy from events that might or might not occur regardless of our beliefs. I will generally label the outcome x.

Risk Specifications h: the family of probability distributions presumed feasible for x. Each member of h is presumed to have a different mean θ, with exactly one $h(\cdot|\theta)$ applying at any given time.

Hidden Information: We never directly observe θ.

True [Mean] θ: the unknown current mean of x. In a Bernoulli $\{0,1\}$ game, where $h(x=1|\theta)=\theta$, I will sometimes call it true risk.

Beliefs: the credence given to various hypotheses about the true θ. Beliefs are sufficiently coherent that they never induce knowingly bad bets. Beliefs can be implicit, with key aspects inferred from bets even when credence is never directly queried.

Subjective Probability p: a quantification of beliefs that satisfies the basic rules of probability. Usually p is continuous with one or two humps, which mark our leading hypotheses about θ. That still allows p infinitely more variety than θ. Figure 1.1 displays a few examples.

Mean Belief E: the expectation $\langle\theta\rangle \equiv \langle\theta\rangle_p \equiv \int \theta p(\theta) d\theta$. It can be inferred from the maximum acceptable bet on x or from the report for a strictly proper scoring rule $S(E,x)$.

Certainty about risk: full confidence that $\theta = E$. It implies a singleton distribution for p, where all weight goes on a single point. I will describe this with the shorthand $p(E)=1$.

Uncertainty about risk: a non-singleton p, or more broadly a shorthand for subjective probability.

Figure 1.1: Possible uncertainties $p(\cdot)$ about true risk θ

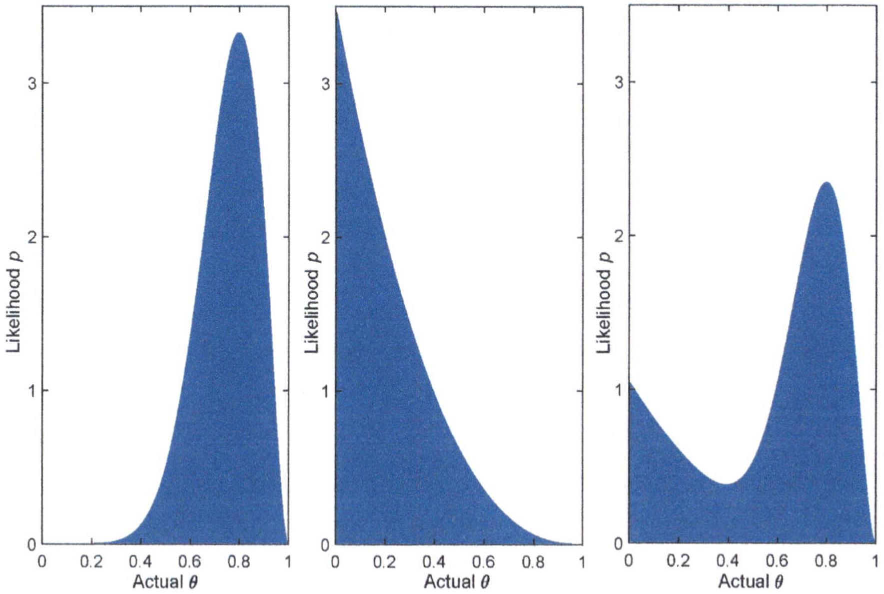

Belief Variance V: the expectation $\langle(\theta-E)^2\rangle=\langle\theta^2\rangle-E^2$, which is positive for any non-singleton $p(\cdot)$. We will learn ways to infer this from two bets.

Certain uncertainty: a fully specified p. This is the standard assumption used to apply Bayes' Rule:

$$p(\theta|x) \propto h(x|\theta)\cdot p(\theta). \tag{1.1}$$

Fuzzy or uncertain uncertainty: a partial specification of p, say E and V only or E with some restricted range for θ.

The most notable objection to subjective probability concerns its empirical fuzziness. Few of the reports we get from others probe much beyond their E. Even our own p is fuzzy. When needed, we conduct thought experiments that sample our current willingness to pay and use that to refine our priors. Fortunately, that doesn't rule out treatment of p as a useful mathematical abstraction, akin to our treatment of unverifiable risk θ or geometric abstractions of points, lines, and surfaces.

After two centuries of derision, subjective probability is now widely accepted. It proved its worth once computers advanced enough to apply it to complex problems. Modern search engines make billions of decisions daily that bridge objective and subjective probabilities. However, there is still some tendency to reduce subjective p to its mean E.

The problem originated when defenders of subjective probability, like Ramsey (1926) and Anscombe and Aumann (1963) equated p for an event to fair price E for a bet paying 1 if the event happens and 0 if it doesn't. This conveyed much of the intuition for subjective probability and helped persuade skeptics. But all the bet really points to is an equivalence class embracing all p with mean E. If beliefs were trimmed to the singleton $p(E)=1$, the next bet involving the same risk would remain E with no learning from experience.

Savage (1954) was more attentive to the differences although his language was loose. The following quotation can be translated into our terms as "$E=\frac{1}{2}$ does not require $p(\frac{1}{2})=1$ or rule out $p(0)=p(1)=\frac{1}{2}$":

> If a person is 'sure' that the probability of heads on the first toss of a certain penny is ½, it does not at all follow that he considers the coin fair. He might, to take an extreme example, be convinced that the penny is a trick one that always falls heads or always falls tails. (p. 24)

Savage also noted that beliefs differ in relative certainty and that uncertain beliefs tended to change faster than uncertain ones. In fact, the measurement of change is key to illuminating relative certainty. Savage never managed to clarify this, to his own chagrin:

> There seem to be some probability relations about which we feel relatively 'sure' as compared with others. When our opinions, as reflected in real or envisaged action, are inconsistent, we sacrifice the unsure opinions to the sure ones. The notion of 'sure' and 'unsure' introduced here is vague, and my complaint is precisely that neither the theory of personal probability, as it is developed in this book, nor any other device known to me, renders the notion less vague. (pp. 67-68)

De Finetti (1979, refining views first published in 1931 and 1937) acknowledged that the E revealed by small bets or proper scoring rules is an expectation. He called it "prevision" and explained in a footnote that

> This corresponds to 'mathematical expectation' in classical terminology, and to 'mean value' in more up-to-date usage. We prefer to reserve the term 'mean value' for objective distributions (e.g. statistical distributions). (p. 64)

When the prevision concerned the occurrence of an event, de Finetti said that it could also be called probability, but he expressed concern about loose wording and offered as a rigorous definition (with my renaming the event x to avoid confusion):

> The probability $P(x)$ that You attribute to an event x is, therefore, the certain gain p that You judge equivalent to a unit gain conditional on the occurrence of x. (p. 64)

At first glance, this seems to identify probability p with a single expectation. However, de Finetti denied any objective risks for p to map, as in his view "only subjective probabilities exist" (p. 3). Instead, he modeled a P that maps from information to fair bets. Consider Savage's two scenarios for the toss of a penny. De Finetti would describe any $E = \frac{1}{2}$ on the first toss x_1 as $P(x_1 = 1) = \frac{1}{2}$ but distinguish fair and trick coins by expectations $P(x_2|x_1)$ for the second toss x_2. In one case, $P(x_2 = 1|x_1) = \frac{1}{2}$ regardless of x_1. In the other, $P(x_2 = 1|x_1) = x_1$.

I applaud de Finetti's focus on responses to additional information. It is the only way to distinguish beliefs with the same E. By defining conditional P without reference to θ, he also shifted attention from degrees of uncertainty to shifts in certainty-equivalent bets. Yet the two are closely related. Uncertainty makes bets shift in response to new evidence while the shifts help reveal the dispersion of beliefs. To highlight the connection, I will stick with the formulation of p as a function of an objective risk θ. The expression $p(\theta) = \frac{1}{2}$ is simpler than $P(x_2 = 1|x_1) = \frac{1}{2}$ and makes the analysis easier to follow.

Measuring Dispersion

Taking multiple bets on a single outcome does not measure the dispersion of beliefs even when x takes a broad range of values. Suppose we offer the proper scoring rule $\mathrm{T}-\sum_{k=1}^{n}|c_k|(x^k-M_k)^2$ to elicit the first n moments $M_1,\dots,M_n$ or $\mathrm{T}+|c|\cdot\log h(x)$ to elicit a full distribution h. Although more expectations are revealed than the scalar report $E=M_1$, they all indicate mean beliefs, not the dispersions around the means.

To infer the variance V of beliefs, we need fair bets on two outcomes. The simplest case is Bernoulli $\{0,1\}$ risk. When the first outcome $x=1$, $p(\theta)$ shifts to $\theta p(\theta)/E$ after Bayesian update. The next bet $E+\Delta E$ is $\langle\theta^2\rangle/E$, so $\Delta E=V/E$. When $x=0$, $p(\theta)$ shifts to $(1-\theta)p(\theta)/(1-E)$, which implies $\Delta E=-V/(1-E)$. Hence

$$V=\begin{cases} E\cdot\Delta E & \text{for } x=1 \\ (E-1)\Delta E & \text{for } x=0 \end{cases}. \tag{1.2}$$

While V does not fully measure the dispersion, it is the best single statistic and makes the vagueness Savage lamented considerably less vague. It also supports Savage's intuition that vaguer beliefs are less stable, as higher V induces higher $|\Delta E|$ all else being equal.

The following generalization of (1.2) is more intuitively appealing and offers more insight into how it works. For any binary outcomes $\{L,H\}$, let $\sigma_E^2=(H-E)(E-L)$ denote the variance of x given E. Then

$$\Delta E=\frac{V}{\sigma_E^2}(x-E) \quad \text{and} \quad V=\frac{\Delta E}{x-E}\sigma_E^2. \tag{1.3}$$

Here $x-E$ denotes the news in x relative to what was expected. The fraction of news incorporated into the next bet equals the ratio V/σ_E^2 of two perceived variances: one for θ and the other for x. Hence Bayesian updating changes bets by the news times the relative variance.

An alternative formulation invokes the correlation ρ between two otherwise independent events driven by the same θ. As the chance that both occur is θ^2, their covariance $\langle\theta^2\rangle-\langle\theta\rangle^2=V$ and $\rho=V/\sigma_E^2$. Hence

$$\Delta E=\rho(x-E) \quad \text{and} \quad V=\rho\sigma_E^2. \tag{1.4}$$

To be clear, this correlation is not inherent in risk; rather, it is induced by our imperfect information p. However, it deserves attention since it implies a likely clustering of outcomes above or below E. Der Kiureghian and Detlevsen (2009) emphasized the importance for engineering safety:

> Uncertainties are characterized as epistemic ["objective risk"] if the modeler sees a possibility to reduce them by gathering more data or by refining models. Uncertainties are categorized as aleatory ["subjective"] if the modeler does not foresee the possibility of reducing them... Most importantly, epistemic uncertainties may introduce dependence among random events.

For an illustration of (1.4), consider a conjugate beta(α,β) prior on Bernoulli risk. Here

$$E=\alpha/(\alpha+\beta), \quad \rho=1/(\alpha+\beta+1), \quad V=\rho E(1-E). \qquad (1.5)$$

After x is observed, p shifts to beta($\alpha+1,\beta$) if $x=1$ or beta($\alpha,\beta+1$) if $x=0$. This implies $E+\Delta E=\rho(\alpha+x)$ and $\Delta E=\rho(x-E)$.

Treating observations as conditionally independent likens inference to analysis of a compound lottery. First θ is selected randomly with known distribution p but not revealed. Then various Bernoulli outcomes with risk θ are observed. Every permutation of k occurrences out of m has probability $q_{m,k}-\langle\theta^k(1-\theta)^{m-k}\rangle$. E reveals only $\{q_{1,k}\}$, while E and V reveal $\{q_{2,k}\}$. E and V do not reveal any higher moments unless $V=0$.

E and V usually leave a lot of wiggle room for the tails. Consider a coin with perceived $E=0.5$ and $V=0.05$. The perceived chance $q_{2,2}$ of two heads in two tosses is 0.3. What is the perceived chance $q_{8,8}$ of eight heads in eight tosses? Let's estimate this using various methods:

Method 1: Assuming a beta distribution, E and V require $\alpha=\beta=2$, in which case $q_{8,8}=0.0545$.

Method 2: Given $\sigma_E^2=0.25$ and $\rho=V/\sigma_E^2=0.2$, eight tosses have mean 4 and standard deviation 2.19, which makes 8 heads a 1.83 standard deviation outlier. Applying a normal approximation, $q_{8,8}=0.0339$.

Method 3: Among binary beliefs p consistent with E and V, the maximum weight on $\theta=1$ is $1/6$ and is coupled with weight $5/6$ on $\theta=0.4$. This implies $q_{8,8}=0.1672$.

Method 4: By symmetry, another belief sets $p(0)=1/6$ and $p(0.6)=5/6$, which implies $q_{8,8}=0.0140$.

Method 5: Taking a convex combination of the binary beliefs in Methods 3 and 4 allows any $q_{8,8}$ between 0.0140 and 0.1672.

Returning to the general case, imagine writing down all the $q_{m,k}$ and lining them up analogously to the Pascal triangle of binomial coefficients $C_{m,k}\equiv m!/k!/(m-k)!$. Like the Pascal triangle, each element sums the two closest neighbors on a neighboring row. However, the flow is up rather than down: $q_{m,k}=q_{m+1,k}+q_{m+1,k+1}$. If we know all values on one row, we can uniquely fill all values above. We can do the same if we know one element on each row. Below I chart the first four rows on the left and fill in some of the elements to its right.

$$
\begin{array}{ccccccc}
 & & & q_{0,0} & & & \\
 & & q_{1,0} & & q_{1,1} & & \\
 & q_{2,0} & & q_{2,1} & & q_{2,2} & \\
q_{3,0} & & q_{3,1} & & q_{3,2} & & q_{3,3} \\
\cdots & \cdots & \cdots & \cdots & \cdots & &
\end{array}
\qquad
\begin{array}{ccccccc}
 & & & 1 & & & \\
 & & 1-E & & E & & \\
 & q_{2,0} & & q_{2,1} & & E^2+V & \\
q_{3,0} & & q_{3,1} & & q_{3,2} & & q_{3,3} \\
\cdots & \cdots & \cdots & \cdots & \cdots & &
\end{array}
$$

Here $q_{2,1}=E(1-E)-V$ and $q_{2,0}=(1-E)^2+V$. Eliciting bets E on the next m outcomes will fill the triangle down to row m. We can gather this information at the beginning or ask m times about the next outcome.

The expectation $p_{m,k}$ of k occurrences out of m in any order equals $C_{m,k}q_{m,k}$. The set $\{\tilde{p}:\tilde{p}(k/m)=p_{m,k}\}$ provides a discrete approximation of p that improves as m increases.[2] The right edge with $p_{m,m}=q_{m,m}$ eventually reveals the moment-generating function $\langle e^{\theta s}\rangle=\sum_{m=1}^{\infty}(p_{m,m}s^m/m!)$ that fully defines p.

The most interesting special case sets $V=0$ as it makes p a singleton. In principle, every objective risk θ can be redefined as a singleton belief with $p_{m,m} = \theta^m$ and all other beliefs expressed as mixtures of singletons. That implicitly presumes unlimited random samples from the same risk.

Is there some intermediate framework that allows limited replicability? Yes, there is. It is called finite exchangeability. For a simple example, consider a contest between three firms to set a new broadband standard, with equal expectations of a single firm being chosen and a consortium of all three firms. Here $q_{3,1} = q_{3,3} = \frac{1}{4}$ and $q_{3,0} = q_{3,2} = 0$. The corresponding triangle is charted below. Its first three rows match those for independent flips of a fair coin tosses. Yet the observations cannot be independent. We cannot add even one more row, since $q_{4,1}$ or $q_{4,2}$ must be positive to sum to $q_{3,1} = \frac{1}{4}$ and this would turn $q_{3,0}$ or $q_{3,2}$ positive. No mixture of Bernoulli risks can account for this.

$$\begin{array}{c} 1 \\ \tfrac{1}{2} \quad \tfrac{1}{2} \\ \tfrac{1}{4} \quad \tfrac{1}{4} \quad \tfrac{1}{4} \\ 0 \quad \tfrac{1}{4} \quad 0 \quad \tfrac{1}{4} \end{array} \qquad\qquad \begin{array}{c} 1 \\ \tfrac{1}{2} \quad \tfrac{1}{2} \\ \tfrac{1}{4} \quad \tfrac{1}{4} \quad \tfrac{1}{4} \\ 0 \quad \tfrac{1}{4} \quad 0 \quad \tfrac{1}{4} \\ 0 \quad ? \quad ? \quad 0 \quad \tfrac{1}{4} \end{array}$$

As de Finetti showed, finite exchangeability of order M is analogous to random sampling without replacement from an urn of M balls. Suppose only red balls are valued at 1. Then each $p_{M,k}$ equals $P(k$ of M are red), to use de Finetti's notation for expectations. This would reinterpret the previous example as P(1 of 3 is red)=¾ and P(3 of 3 are red)=¼.

Finite exchangeability is a clever way to quantify a fuzzy uncertainty without requiring its extension to a fully specified p. However, it is more a twist on conditional independence than a replacement for it. We still treat balls in the urn as equally likely to be picked at any given round.

More importantly, it doesn't address the core challenge of ambiguity. Few decisions are fully replicable even once. When they aren't, de Finetti's logic allows estimation of E only, as if degrees of conviction don't matter. But they do. They influence how much new information will change E, which in turn influences how eager we are to gather it.

For example, suppose snow next Saturday, whose chance is E judging from Monday's forecast, would interfere with a couple's planned outdoor wedding. Moving to an indoor venue is possible but the costs rise the longer the delay. If wholly confident about E, the couple should decide immediately. However, since the forecast will likely be refined over the next few days and drive E towards 0 or 1, perhaps the couple should wait to decide. Their what-if decisions would help reveal their uncertainty.

Uncertainty about E is central to this book. Throughout I take a utilitarian approach, where ambiguity per se doesn't affect decisions. In decision theory this is known informally as the "sure thing principle" and formally as convexity. It states that if an action is preferred whether the relevant probability distribution is p or q, no uncertainty between p and q will affect it.

Choices must be convex to be consistent with expected utility theory (Green and Osband 1991, Osband 2023). While violations are common enough to inspire a huge literature, I will mostly ignore them.[3] Every violation points to a series of bets that the violator will willingly accept but lose money on (Green 1987) and finance abounds with traders looking to exploit that. This generates huge incentives to pay attention, to gauge tradeoffs more consistently, or to transfer wealth to those who do.

To be clear, I do not claim that financial markets are superbly prescient. Their forte is error correction, not prophecy. Practicalities favor simpler hypotheses with fewer variables and some scenarios ignored or misunderstood. As we will see, differences in interpretation are key to understanding the huge volume of trading and the extra risk it generates.

However, professional traders should not be treated as no better at handling huge bets than students in lab experiments are at handling small bets. In 15 years of working with experienced traders at investment banks and hedge funds, originally as an economist and later as a portfolio strategist, I found them far shrewder and calmer about markets than I was, albeit with much shorter-term focus. That realization inspired the long investigations summarized in this book.

2

Benefits of Variance

All scientific knowledge is uncertain. This experience with doubt and uncertainty is important. I believe that it is of very great value, and one that extends beyond the sciences. I believe that to solve any problem that has never been solved before, you have to leave the door to the unknown ajar. You have to permit the possibility that you do not have it exactly right. (Feynman 1955)

The more dispersed our beliefs, the more confused we feel about the risks. No one likes confusion. We want clarity, with beliefs tightly clustered around the true θ. But tight clustering around E doesn't guarantee that. Sometimes θ isn't close to E and occasionally it lies far away. When that occurs, tight clustering impedes correction. It conserves large errors far too long. This recalls Mark Twain's alleged quip: "It ain't what you don't know that gets you into trouble. It's what you know for sure that just ain't so." While certainty about θ is our theoretical ideal, in practice it helps to retain some doubt.

From (1.3), the response ΔE to any binary outcome x equals the relative variance V/σ_E^2 times the news $x-E$. To extend this result, suppose the outcome dx over an instant dt is $\pm\sigma\sqrt{dt}$ with true expected value $\mu\,dt$. Effectively this is diffusive Brownian motion with drift μ and volatility σ. If beliefs about μ have mean E and variance V, then dE equals V/σ^2 times $dx-E\,dt$. Hence binary and Brownian outcomes have similar effects when σ is known:

$$dE = \frac{V}{\sigma^2}\left(dx - Edt\right) = \frac{V}{\sigma}dy \quad \text{for} \quad dy \equiv \frac{dx - Edt}{\sigma}. \tag{2.1}$$

Since dy is standardized Brownian noise, dE has volatility V/σ, just as ΔE has standard deviation V/σ_E. Suppose the true or μ shifts dramatically: say, a new trend rate of economic growth. Then higher V induces both quicker and noisier adjustment. In short, there is an inherent tension between stable know-how and dynamic learning. If we think of learning fitness in an evolutionary sense, low V facilitates fine tuning while high V provides more robustness.

In fact, there is a close analogy between learning tradeoffs and the tradeoffs in evolutionary biology. While Darwin (1859) emphasized both survival of the fittest and the variety of adaptation, early efforts at formalization could not reconcile the two. Why shouldn't natural selection filter for the best single types and extinguish the rest? The puzzle was heightened by Mendel's discovery that preservation of genetic variety was more the rule than the exception.

Fisher (1930) provided the first formal resolution, which he called the fundamental theorem of natural selection and stated as follows: "The rate of increase in fitness of any organism at any time is equal to its genetic variance in fitness at the time". To understand its gist, consider two competing genes in proportions F and $1-F$. Suppose that the first gene outcompetes the second at rate α. After an instant dt,

$$F + dF \approx \frac{F + \alpha F dt}{1 + \alpha F dt} \quad \Rightarrow \quad dF = \alpha F\left(1 - F\right)dt. \tag{2.2}$$

The mean fitness advantage of the gene pool is $E = \alpha F$ while its variance is $V = \alpha^2 F(1-F)$, so (2.2) implies $dE = Vdt$, which echoes (2.1) apart from the transformation of repeated x into a deterministic trend.

Despite Fisher's esteem as both evolutionary biologist and statistician, his theorem met with a mixed reception. It seemed confusing and was often misunderstood. Here it is useful to distinguish two main kinds of fitness. Static fitness favors survival in a stable environment. Dynamic fitness favors adaptation to changing environments. Fisher emphasized that environments evolve too, largely because statically fit species exploit them, which makes dynamic fitness more important in the long run. Dynamic fitness is better known as robustness. Taleb (2012) popularized a similar notion under the name of "antifragility". Chapter 4 will reveal additional kinds of fitness, which facilitate shifts from refinement to robustness and back.

The main caveat about (2.2) is that it drives F toward the boundaries where V vanishes. Similarly, V in (2.1) usually dwindles as the number T of observations grows. To keep the dynamics interesting, θ or μ must occasionally change, with (2.1) modified to allow some anticipation of change. As the next chapter will show, even tiny doubts about stability can induce major, complex changes in V. Why does that matter? The core answer is that V signals the value of information.

To see that, let's return to the strictly proper scoring rules $S(E,x)$ mentioned in the previous chapter. Define $R(E)=S(E,E)$. Then S will be strictly proper if and only if R is strictly convex, with general formula

$$S(E,x)=R(E)-R'(E)(E-x)+\mathrm{T}(x), \tag{2.3}$$

where T is arbitrary and R' denotes any subgradient at kinks (Osband 1985, Gneiting and Rafferty 2007). Usually scoring rules are posed as if the agent knows the true distribution of x, so that the perceived E matches the true mean θ. In that case, setting $R(E)=|\varepsilon|E^2$ for miniscule ε suffices to induce truthfulness. Selection of S becomes far more interesting when an uncertain agent needs to be incentivized to learn.

Ideally the agent might deduce some new scientific law or previously unnoticed variable z that fully identifies θ. However, discovery will likely depend far less on a particular scoring rule than on a broader research agenda. The point of S is to encourage closer attention to estimation using techniques already at hand.

Accordingly, I will model refinement as a thought experiment that enables a Bayesian update $\Delta E=V(x-E)/\sigma_E^2$ without risking a change in θ. This raises the expected return for truthful reporting by

$$R(E+\Delta E)-R(E)\approx R'(E)\Delta E+\tfrac{1}{2}R''(E)(\Delta E)^2 \tag{2.4}$$

and hence raises the expected score by

$$\Delta\langle S(E,\theta)\rangle \approx \tfrac{1}{2}R''(\theta)\langle(\Delta E)^2\rangle = \tfrac{1}{2}R''(E)\cdot\frac{V^2}{\sigma_E^2}. \tag{2.5}$$

If observation costs c per unit of observation variance, then investigation is warranted until V shrinks to approximately

$$V^* \approx \sigma_E^2\sqrt{2c/R''(E)}\,. \tag{2.6}$$

The higher $R''(E)$, the higher V, or the lower σ_E^2, the more likely refinement is justified. Due to the quadratic dependence in (2.5), incentives rise fast as V rises and drop fast as V falls. For quadratic R, incentives are independent of E. For entropic measures $R''(E) \propto 1/E$ and hence encourages extra attention to the prediction of rare events.

Uncertainty in Machine Learning

The most common operation in machine learning resembles a Bayesian update. It refines an estimate E by nudging it a small fraction η towards the latest observation x:

$$\Delta E \approx \eta(x-E). \tag{2.7}$$

η is called the learning rate. For Bernoulli and Brownian risk, as we saw in (1.3), (1.4) and (2.1), the optimal Bayesian learning rate is given by the variance ratio V/σ_E^2, or equivalently by the correlation ρ between two conditionally independent outcomes:

$$\eta = V\big/\sigma_E^2 = \rho\,. \tag{2.8}$$

While machine learning algorithms rarely adjust η explicitly for the scale σ_E of noise, they achieve the same core effect by standardizing the input data $x-E$. The occasional shrinkage of η is broadly consistent with the dwindling of V as beliefs become more certain. While η is typically much smaller than V, this can be explained by frequent iteration through the same data. Each round extracts only part of the useful information, which wastes time in estimating a scalar but helps smooth multivariate estimation when the covariance matrix Σ of noise is ill-conditioned. For example, the popular Adam method (Kingma and Ba 2015) adjusts endogenously for σ_E^2 using an exponentially weighted moving average of squared deviations. It then mildly smooths each $x-E$ estimate (which adds momentum) before applying a small core multiplier.

Most multi-layered neural networks are trained without much effort to endogenously estimate V. One reason is the extra computation it requires. Another is skepticism about the usefulness. However, without any tracking of V or ρ, machine learners tend to spend too much effort trying to categorize inherently ambiguous data.

Figure 2.1 gives examples from the MNIST data set. The top row depicts some handwritten 4s that are often misidentified as 9, while the bottom row depicts some 9s that are often misidentified as 4. These examples come from the subset for testing and hence are not explicitly used for training. However, they have indirect influence as algorithms are typically refined and filtered for those that fit the test set best.

If we score an actual 9 as 1 and an actual 4 as 0, then the difficulty of identifying the symbols in Figure 2.1 suggests that each E is much closer to ½ than to 0 or 1. Their writers almost surely had one of the two in mind, and our training methods push beliefs toward the edges. Hence E can only

Figure 2.1: Ambiguous 4s (top row) and 9s (bottom row) from MNIST test data

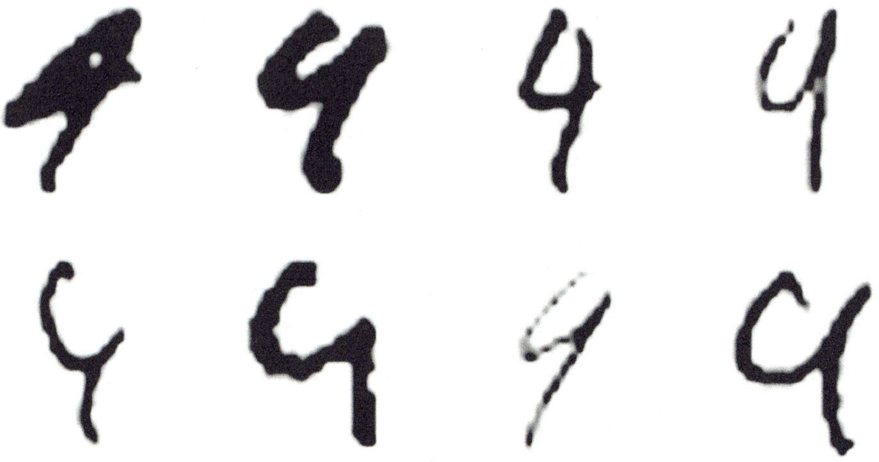

stay intermediate if the relevant V/σ_E^2 is quite small. It is as if the neural network identified a rare, borderline 4-or-9 type.

Awareness of ambiguity can usefully motivate us to look for auxiliary data. Perhaps we can find a data set so massive that it provides many examples of borderline 4-or-9. In Figure 2.1 the 9s are slightly more curved than the 4s; the neural network might learn to recognize that feature and invoke it. It is likely even more useful to look at neighboring digits for clues about writing style. If it is part of an address, then compare candidates with other information on location.

Sometimes machine learning presses for too much certainty too quickly, with embarrassing or harmful results. To its credit, the machine learning community frankly acknowledges the dangers. As noted by Kendall and Gal (2017) in an article widely cited since,

> [Deep learning] mappings are often taken blindly and assumed to be accurate, which is not always the case. In two recent examples this has had disastrous consequences. In May 2016 there was the first fatality from an assisted driving system, caused by the perception system confusing the white side of a trailer for bright sky. In a second recent example, an image classification system erroneously identified two African Americans as gorillas, raising concerns of racial discrimination. If both these algorithms were able to assign a high level of uncertainty to their erroneous predictions, then the system may have been able to make better decisions and likely avoid disaster.

On the other hand, occasionally we can become so confident about our uncertainty that it is not worth resolving further. If E seems to stabilize apart from sample noise, repeating the same kinds of training isn't likely to help. In reinforcement learning, machines seeking to design optimal policies continually make tradeoffs between exploiting what they already know and experimenting to learn something new. Hence, as Osband I. et al (2021, 2022) emphasize, learning efficiency can be greatly improved by distinguishing two scenarios:

- High subjective V relative to objective σ_E^2, aka high ρ, aka high epistemic uncertainty relative to aleatoric uncertainty.
- High objective σ_E^2 relative to subjective V, aka low ρ, aka high aleatoric uncertainty relative to epistemic uncertainty.

From (1.3) and (2.1), the former deserves far more attention than the latter.

A classic image in Figure 2.2 illustrates the challenge well. It resembles either a duck or a rabbit, depending on the perspective. There is not much point to learning which of those two classifications it most deserves.

The emergence of ChatGPT and other AI experts makes it more important than ever to track uncertainty and report it. They are far more versatile than human minds and occasionally surpass human experts. But they are prone to present consensus human opinions as incontrovertible truths, even when consensus is defined only by internet ratings of popularity or by their creators' personal rankings. They also extend their inferences into areas they can't understand and hallucinate supposed facts. AI experts will be wiser when they better acknowledge what neither they nor their information sources know.

Figure 2.2: Rabbit or duck?[4]

Uncertainty in Finance

From (1.3), the dispersion of beliefs is often best probed by tracking the changes in bets they induce. Hence financial markets offer great windows into uncertainty. Unfortunately, the two macroeconomists who most highlighted uncertainty ruled out its quantification. Knight (1921) defined uncertainty as unmeasurable risk:

> A *measurable* uncertainty, or "risk" proper, as we shall use the term, is so far different from an *unmeasurable* one that it is not in effect an uncertainty at all. We shall accordingly restrict the term "uncertainty" to cases of the non-quantitative type. It is this "true" uncertainty, and not risk [...] which forms the basis of a valid theory of profit. (Chapter 1)
>
> Take as an illustration any typical business decision... What is the "probability" of error (strictly, of any assigned degree of error) in the judgment? It is manifestly meaningless to speak of either calculating such a probability a priori or of determining it empirically by studying a large number of instances... [T]he "instance" in question is so entirely unique that there are no others [...] to form a basis for any inference of value about any real probability in the case we are interested in. (Chapter 7)

Similarly, Keynes (1921) claimed that only beliefs grounded in solid statistical evidence or logical assumptions (e.g., equal likelihoods of base-level events) deserved probabilistic quantification; subjective beliefs about unique events did not. His later writings explicitly placed most financial forecasts in the latter camp:

> The outstanding fact is the extreme precariousness of the basis of knowledge on which our estimates of prospective yield have to be made.... Our existing knowledge does not provide a sufficient basis for a calculated mathematical expectation. (Keynes 1936, Chapter 12)

> By "uncertain" knowledge, let me explain, I do not mean merely to distinguish what is known for certain from what is only probable. The game of roulette is not subject, in this sense, to uncertainty... Even the weather is only moderately uncertain. The sense in which I am using the term is that in which the prospect of a European war is uncertain, or the price of copper and the rate of interest twenty years hence... About these matters there is no scientific basis on which to form any calculable probability whatever. We simply do not know. (Keynes 1937)

In short, Knight and Keynes thought that key investment risks were too complex and unique to reasonably quantify. They were only dimly aware of subjective probability and missed the connection between E and market prices. The best that capital market pricing can do is balance supply with demand. In Keynes' view, that was such a big challenge that it warranted more fiscal and monetary intervention. But he was not looking for a fair value anchor. He did not think there was one.

By the 1960s, subjective probability was more accepted in economics. In a now-classic article, Samuelson (1965) contended that efficient capital markets incorporate all available information, including agents' beliefs about the future. However, he did not claim that markets predicted the future well and explicitly acknowledged a host of new questions:

> I have not here discussed where the basic probability distributions are supposed to come from. In whose minds are they ex ante? Is there any ex post validation of them? Are they supposed to belong to the market as a whole? And what does that mean? Are they supposed to belong to the 'representative individual', and who is he? Are they some defensible or necessitous compromise of divergent expectation patterns? Do price quotations somehow produce a Pareto-optimal configuration of ex ante subjective probabilities? This paper has not attempted to pronounce on these interesting questions.

The next generation of theorists focused on the neatest subset of these questions: expectations that accurately gauge future risks. Their justification is folded into the name Rational Expectations. Rational economic agents have incentives to tune their beliefs to the true risk θ. Their beliefs should not be systematically biased, so $E=\theta$ is a reasonable baseline. Furthermore, beliefs that deviate from θ will eventually be punished by market outcomes. Rational Expectations implicitly assumes that enough relevant data accumulates to ignore V. Formally it needs to ensure this through full information, hence the expanded acronym FIRE.

FIRE's insights and tractability have made it the dominant paradigm in macroeconomics and finance. However, some market behavior cannot be reconciled with FIRE. The two best-known examples are the excess volatility of equity prices relative to dividends and the volatility of excess volatility. Shiller (1981, 1989, 2000) attributed excess volatility to irrational exuberance. Engle (1982, 2001) developed models of Generalized Autoregressive Conditional Heteroskedasticity (GARCH) that describe volatile volatility neatly but have no apparent rationale. Both Shiller and Engle were awarded Nobel Prizes for their work, testifying to its broad acceptance and importance.

Volatile V can easily explain this. David (1997) showed that rational tracking of shifts between high and low firm profitability induced GARCH behavior. Kurz (1997) attributed 75% of asset return variability to traders' rational uncertainties about each other's demand functions. Kurz (2009) extended this work to explain both GARCH behavior and excess volatility. Pastor and Veronesi's (2009) survey of rational learning reached similar conclusions. Chapter 4 will trace both phenomena to the cumulant hierarchy of learning.

The most flagrant violation of FIRE is the huge volume of trading, which shows that disagreement is rife. Nobel Prize winners Hansen (2007, 2010), Sargent (2008), and Stiglitz (2010) each acknowledged the need to incorporate heterogeneous beliefs. Yet Kurz (1994ab, 1997) had already developed a theory of "rational beliefs" that allowed for heterogeneity. Why didn't this attract more attention from the profession? One clue comes from Dumas et al (2009), who claimed that "heterogeneity of beliefs between agents needs to be regenerated as otherwise Bayes' law causes it to die out". That is true but misleading, as it suggests that regeneration must be supplied externally. Chapter 6 will show that rational turbulence naturally regenerates heterogeneity.

Unfortunately, much of the literature on capital markets continues to treat only behavior that conforms to FIRE as truly rational. When new evidence rolls in, it is considered to confirm either FIRE on how investors should act or Behavioral Finance on how foolishly they do act. Each school basks in its supporting evidence and inclines to ignore the rest. This impedes new thinking. Imagine how slower quantum theory would have developed if the wave and particle behaviors it sought to reconcile had been attributed to two different kinds of light.

The aspect of the Rational/Foolish divide that is hardest to explain is why it persists. Why hasn't the Foolish pool smartened up over time? Why haven't Rational traders separated the remaining Fools from their money? Implicitly, there must be some everlasting fountain of foolishness from which traders imbibe. But how could this possibly be evolutionarily stable?

Taleb (2007) offered a candidate explanation and popularized it. He posited two market realms. In orderly 'Mediocristan', which dominates most of the time, ordinary probabilistic inferences apply. In turbulent 'Extremistan', they do not. Taleb related this to human reluctance to acknowledge unpredictability and occasionally dramatic regime change. People exaggerate stability when they can and panic when they cannot.

Empirically, financial markets do appear calm most of the time and very turbulent when they're not. People do seem to swerve from complacency to panic. Still, Taleb's division is only sustainable when Foolishness dominates both sides. Otherwise, rational traders would factor extreme risks into their valuations, bet defensively in Mediocristan and reap huge windfalls in Extremistan. Their bets would then carry much greater weight, which would make the next Mediocristan less mediocre and the next Extremistan less extreme. In a generation or two this would render most of Taleb's claims moot.

Tracing market turbulence to irritated spirits is like tracing turbulent air or water to irritated gods. Granted, it looks like something external is stirring things up, which is why ancients conjured up wind and water spirits. But we now know that turbulent flow just reflects the same fluids moving so fast against resistance that the molecules tumble over each other and swirl. Similarly, we will see that fast learning after big surprises makes market prices turbulent.

Panic is more a reaction to turbulence than its cause. When a new trend emerges in markets, people aren't sure what to make of it. Initially they treat the observations as random outliers. As those outliers accumulate, tiny doubts start to mount, until at some point it isn't clear whether mean beliefs are way too sensitive to news or not sensitive enough. That is when variance most matters.

Looking back, finance theory has grappled with uncertainty for a century without getting a firm grip. Unquantifiable risk is too vague a notion. FIRE is too certain. Foolishness is too averse to learning. The best way forward is to model rational learning about unstable trends.

This need not shatter behavioral finance. The core learning equations are too complex for any intelligence to fully apply. Humans are bound to draw on simpler, intuitive heuristics that are usually effective and occasionally deceptive. Behavioral finance can help identify shortcomings and offer corrections.

A more unified perspective on capital markets can also help improve macroeconomic policy. Central banks and other regulators have become too prone to view market turbulence as foolishness warranting intervention and market calm as wisdom warranting inaction. The former encourages excessive leverage by major speculators as they expect to be bailed out in crises. The latter encourages complacency about mounting sovereign debt whenever interest rates are low. Chapters 9 and 10 will address some of the policy challenges.

3

Tiny Doubts

Frequentist estimates of means simply average the observations. If E denotes the sample mean after $T-1$ observations, the change ΔE after the next observation x is given by

$$\Delta E = \frac{x-E}{T}. \tag{3.1}$$

Let's compare this to the pure Bayesian update of E from (1.3) and (2.1). There the multiplier is V/σ_E^2, which can summarize as the learning rate Hence the reciprocal of η indicates an effective sample size or duration T_{eff}, if all other observations were weighted like the current x:

$$T_{eff} \equiv \frac{1}{\eta} = \frac{\sigma_E^2}{V}. \tag{3.2}$$

For a conjugate beta(α,β) prior, $T_{eff} = \alpha+\beta+1$ as noted in (1.5). It grows by 1 every observation while η continually shrinks. For general p, Bayesian updates (1.1) will eventually drive p toward a beta distribution, so that in the long run T_{eff}/T approaches 1. This is ideal when θ is fixed. $T_{eff} \ll T$ would greatly underweight early information and E would wobble too much. But if θ dramatically changes, high T_{eff} is embarrassing. It anchors E to past observations that are no longer relevant.

To take an extreme example, suppose that 10,050 coin flips are observed and 5,050 are heads, with X heads occurring in the last 50 flips. If heads pay 1 and tails pay 0, what is a fair price E (fair in the martingale sense of zero expected profit) for a bet on the next toss? Standard hypothesis tests accept $E=0.5$ since the sample mean lies only half a standard deviation away. Pure Bayesian E lies between 0.5 and 0.5025. Neither method factors in X. Why not? The classic textbook answer is that only the summary statistics matter. That must be true if, as is usually presumed, the flips are iid. The reference to X might seem an intentional distraction, posed to sift good probability analysts from bad ones. No X between 18 and 32 should alarm.

However, in this case, $X=50$. Not a single tail in the last 50 flips! The chance of that occurring with $\theta=0.5$ is less than 10^{-16}. Almost surely θ has soared. If you can bet on heads for 0.6, do it. But no well-informed observer will take the other side of the bet, apart from a few AI bots clinging to iid. The latter would flunk a Turing test for human judgment, as even 200 straight tails wouldn't seriously shake their confidence.

Tiny doubts have a peculiar impact. When risk seems stable, they hardly matter. When risk clearly changes, they trigger a sharp transition. In the grey area between, doubters can seem to veer from excessive complacency to panicked overreaction.

To investigate, let's assign times $t=1,\cdots,50$ to the last 50 heads and assume that beliefs p just prior to that are beta(100,100).. After $t=30$, pure Bayesian learning updates beliefs to beta(1030,100) with $E=0.565$. While significantly higher than if iid were assumed throughout, it greatly understates how corrosive tiny doubts can be. Suppose we allow a 0.01% chance that our initial beta(100,100) prior ceased to be relevant due to a random beta (1,1) draw from the unit interval. This has no impact on E and raises $\sqrt{V}$ by just 0.0001. Nevertheless, after 30 rounds E reaches 0.968.

As Figure 3.1 shows, the surge is extremely uneven. It takes 15 straight heads for E to exceed 0.55, but 8 heads later E exceeds 0.9. The associated surge in η has two main sources. One is a growing gap after observing T straight heads between the conditional means $E_{0T}=(100+T)/(200+T)$ if θ is constant and $E_{1T}=(1+T)/(2+T)$ if θ has changed. The other is the rising probability weight w_{1T} on a new θ. These two factors overwhelm the declining conditional variances. At every head,

Figure 3.1: Evolution of E with tiny doubt about constant θ

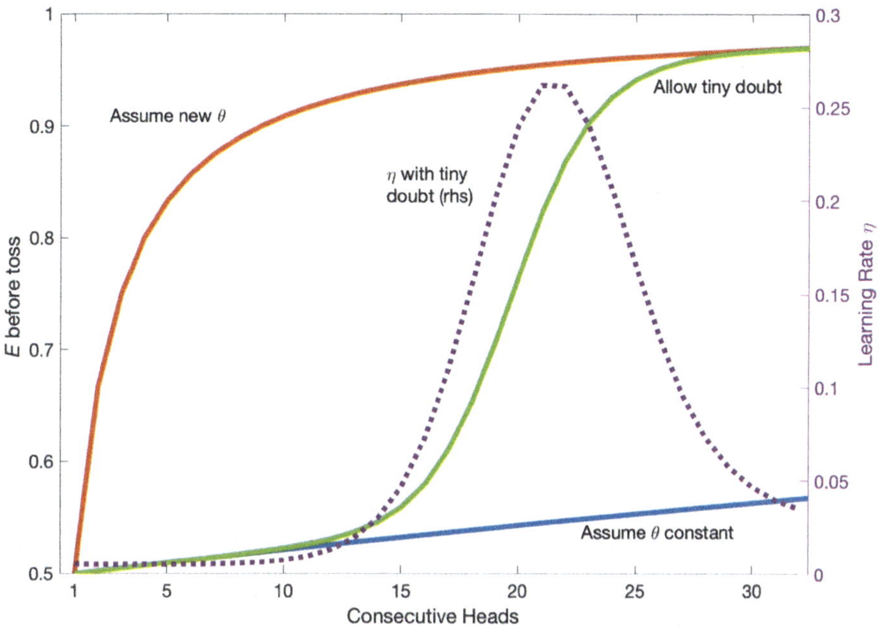

$$
\begin{aligned}
E_T &= \omega_{0T} E_{0T} + \omega_{1T} E_{1T} \\
w_{0,T+1} &= \omega_{0T} E_{0T} / E_T \\
w_{1,T+1} &= \omega_{1T} E_{1T} / E_T
\end{aligned}
\tag{3.3}
$$

The change in $\log(w_{1T}/w_{0T})$ equals $\log(E_{1T}/E_{0T})$, which if perfectly stable describes a logistic curve. The slower E_{0T} moves, the longer E downplays new information but the steeper the core transition. For example, if our baseline p is beta(5000,5000) rather than beta(1000,1000), E arrives at 0.55 one observation later but at 0.90 two observations sooner.

For more flexible regularity, let's model learners who consistently combine Bayesian updates about the past with constant apprehension about future changes. I will call them Tiny Doubters, or TD for short.

As noted previously, the Bayesian update multiplies each belief $p(\theta)$ by θ/E when $x=1$ and by $(1-\theta)/(1-E)$ when $x=0$. We can rewrite the change Δ_b it induces in p as

$$\Delta_b p(\theta) = p(\theta)(\theta - E)\frac{x - E}{\sigma_E^2}. \tag{3.4}$$

TDs follow this with a forward-looking adjustment Δ_f, which anticipates a small chance λ that θ resets to a new value drawn randomly from a beta(α,β) distribution:

$$\Delta_f p(\theta) = \lambda\left(\text{beta}(\theta|\alpha,\beta) - p(\theta) - \Delta_b p(\theta)\right). \tag{3.5}$$

The total adjustment $\Delta = \Delta_b + \Delta_f$ sums (3.4) and (3.5):

$$\Delta p(\theta) = (1-\lambda)p(\theta)(\theta - E)\frac{x - E}{\sigma_E^2} + \lambda\left(\text{beta}(\theta|\alpha,\beta) - p(\theta)\right). \tag{3.6}$$

While we can track the expanding mixture of beta distributions precisely, this quickly becomes unwieldy. My simulations discretize p and update each $p(\theta_i)$ directly using grids of at least 200 points.

To illustrate the diversity of responses, let us warm up different TDs on the same 9,999 flips of a $\theta = 0.5$ coin, followed by one tail at time 0 and 50 straight heads. Let's choose TD α and β from {1,2,3,4,5} and λ from $\{10^{-3}, 10^{-4}, 10^{-5}, 10^{-6}, 10^{-7}\}$. Figure 3.2 plots E for the 125 TDs. While they all respond better than pure beta priors, there is widespread disagreement and no obvious consensus. After 25 straight heads, the simple average E is 0.72 but nearly a quarter of TDs bet on $E < 0.53$ and nearly a quarter bet on $E > 0.9$. How should we choose between them?

The most robust solution treats each $(\alpha_i, \beta_j, \lambda_k)$ as a hypothesis TD_{ijk} about the dynamics of θ with subjective probability $P(\text{TD}_{ijk})$. P gets updated after every observation by applying Bayes' rule to the various E_{ijk} and then factoring in some small chance Λ that the best TD resets. This is wholly analogous to how each TD updates its internal p except that we simplify the reset distribution to uniform, with every TD_{ijk} equally likely to be chosen. Thus we install a TD_{agg} to oversee the base-level TDs and aggregate their E_{ijk} into a consensus E_{agg}.

Figure 3.3 displays E_{agg} for Figure 3.2 with Λ ranging from 10^{-3} to 10^{-8}. The paths are far closer to each other than the TD paths in Figure 3.2 and nearly identical for $\Lambda \leq 10^{-6}$, but some significant differences remain.

Figure 3.2: Evolution of E for various Tiny Doubters

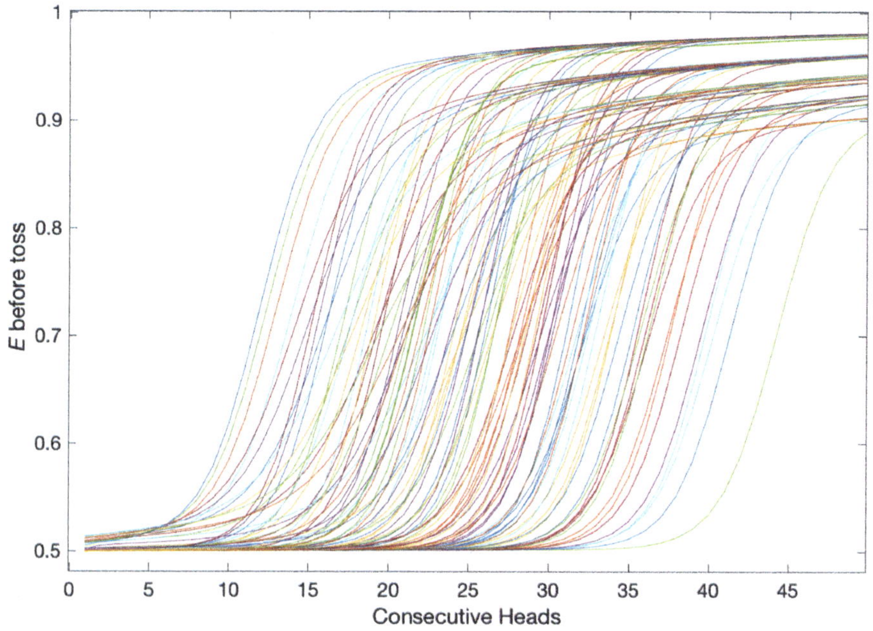

Figure 3.3: Aggregate E for Figure 3.2 using layers of Tiny Doubters

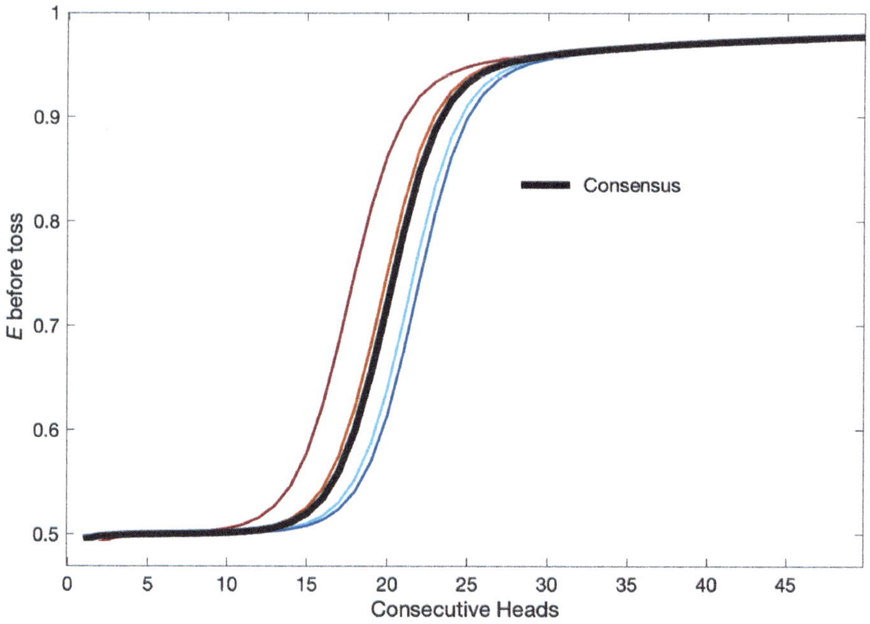

To reconcile, we can set a third layer on top that aggregates various TD_{agg}. Since its values don't vary much with levels of doubt, I call it the consensus predictor. Figure 3.3 displays it as a thick line.

A consensus formed by layers of Tiny Doubters never erases all fuzziness. There are always hypotheses that could be added or modified. However, the variants cluster far closer to each other and look far more quasi-human than predictions that assume perfectly iid observations.

AI experts have started to appreciate this. When asked in February 2025 how many straight heads would raise E to 0.9, their initial answers were 40,000. When prodded about possible coin switch, ChatGPT-4o said 100 to 200 while o3-mini-high said 14 to 28. Grok 3 without prodding rethought its initial answer and said that 20 to 30 should generally suffice.

Perfect Learning

To probe deeper, let's replace one-time shocks with occasional random jumps in θ. Over time, more conviction should accrue to Tiny Doubters that broadly get the transition probabilities right. If one hypothesis fully captures the drivers of θ, eventually it should squeeze all others to the margins. I call it Perfect Learning. To make $TD(\alpha,\beta,\lambda)$ a Perfect Learner, I presume a hidden Markov model with the following properties:

- θ has probability λ at very round of being redrawn from a beta(α,β) distribution and otherwise stays the same.
- A value z is then drawn randomly from the unit interval, with report $x=1$ for $z \leq \theta$ and $x=0$ otherwise.
- α, β and λ are known but θ is never directly observed.

Figure 3.4 displays θ and E for 2000 consecutive plays of Perfect Learning with $(\alpha,\beta,\lambda)=(1,1,0.005)$. Compared with updates of a single conjugate prior, the evolution is strikingly uneven. There are numerous false starts (adjustments to spurious jumps in θ) that later get reversed. Learning prowess manifests mainly in the short lags between big shifts in θ and corresponding adjustments in E. Also note the thinness of most spikes, which indicates that correction is quick.

Figure 3.4: Perfect Learning for 2000 plays with $(\alpha,\beta,\lambda)=(1,1,0.005)$

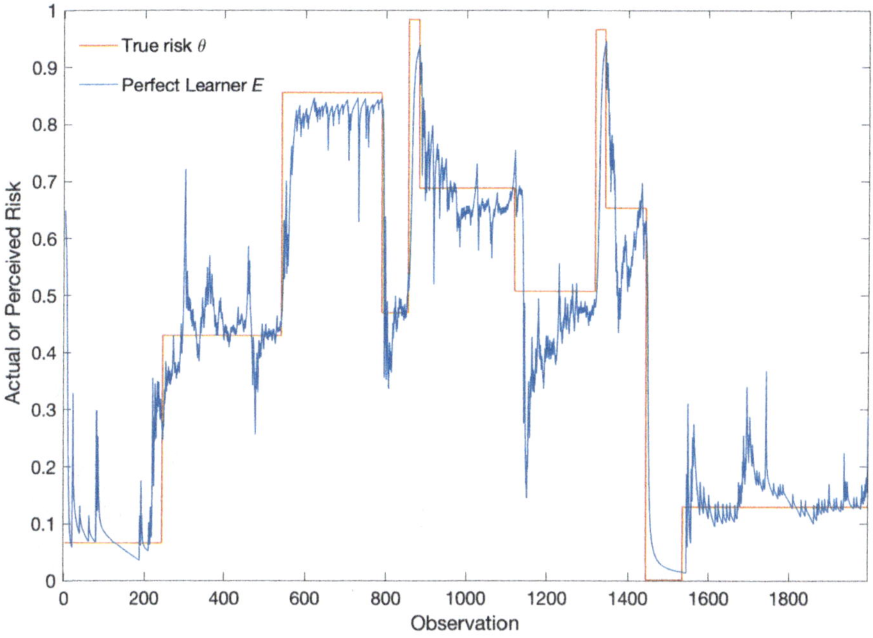

In a million-play simulation, θ changes once every 200 plays on average with an average absolute shift of 0.33. Half of Perfect Learner E lie within 0.03 of the true θ and nearly three-fourths lie within 0.06 of θ. However, 4% of the time E and θ are over 0.2 apart.

Figure 3.5 charts the corresponding $\eta \equiv V/\sigma_E^2$ and reciprocal T_{eff} on a logarithmic scale. T_{eff} never reaches 100 even though the mean time since θ last switched is 200 and many sequences run longer. The long-term average η is 0.052, corresponding to a T_{eff} of 19. The myopic focus reflects the significant chance that older observations have ceased to be relevant. Better to be approximately right than precisely wrong.[5]

The variations in T_{eff} are striking too. Nearly half exceed 30 but 10% are less than 10 and 1.5% fall below 4. Low values tend to cluster in brief spurts, which sometimes shift E to a new plateau and sometimes just add a lot of noise. The spurts themselves are uneven. Figure 3.6 charts the changes ΔT_{eff} from one play to the next.

Figure 3.5: Learning rates and T_{eff} for Figure 3.4

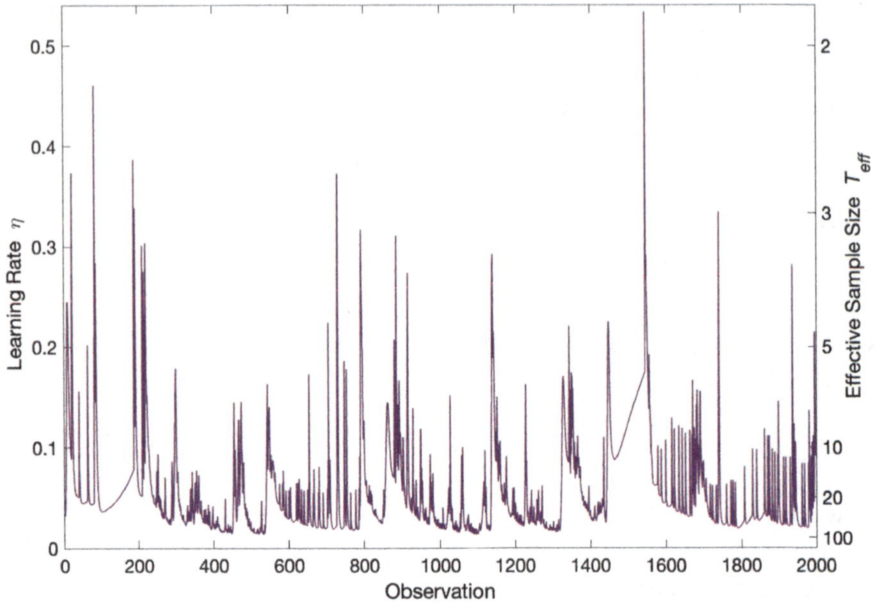

Figure 3.6: Changes in T_{eff} for Figure 3.5

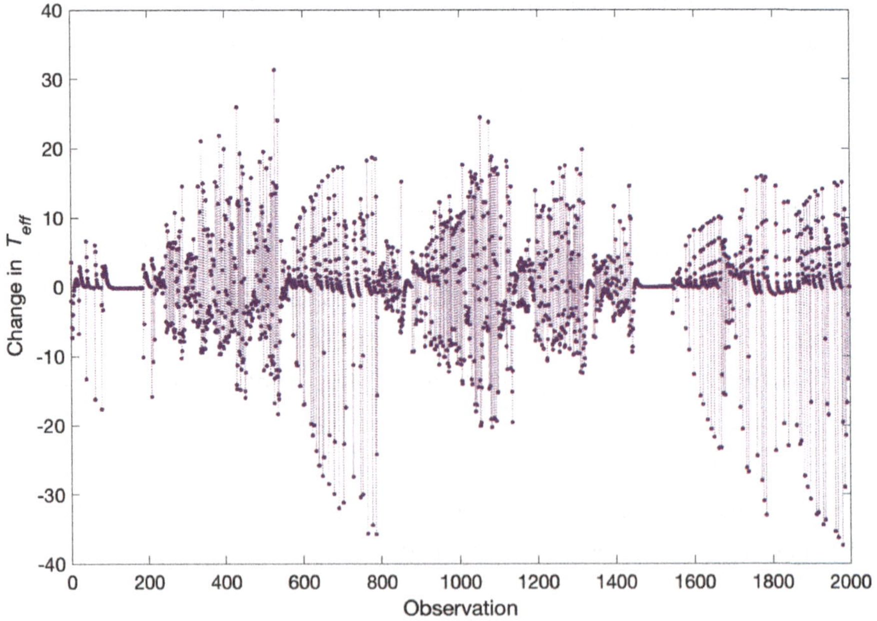

When θ is presumed stable, T_{eff} tends to mount by about 1 every day. But here less than 10% of ΔT_{eff} round to 1. In a million observations, half of absolute daily changes exceed 3.5 and 10% exceed 16. Why weight neighboring x so unequally?

For perspective, let's stabilize η and see what happens. Such estimates are known as exponential moving averages or EMAs. Since θ has λ chance every period of reset, let's start with $\eta = \lambda$. Figure 3.7 graphs its application to our test case. While its E is very smooth compared to Perfect Learner E, its adaptation to new θ is sluggish. Even an EMA with $\eta = 4\lambda$ looks too slow.

When fit is measured by RMSE, the best EMA sets $\eta \approx 0.063$ for $T_{eff} = 16$. Its RMSE is 0.106, compared with 0.134 for $T_{eff} = 50$ and 0.207 for $T_{eff} = 200$. But the best EMA has 20% higher RMSE is 20% higher than the Perfect Learner's 0.089 and its average absolute daily change in E is 40% higher. Hence, even with a fixed η, good tracking favors much faster adjustment to news than the actual λ seems to warrant.

Figure 3.7: Mean beliefs for three EMAs given observations for Figure 3.4

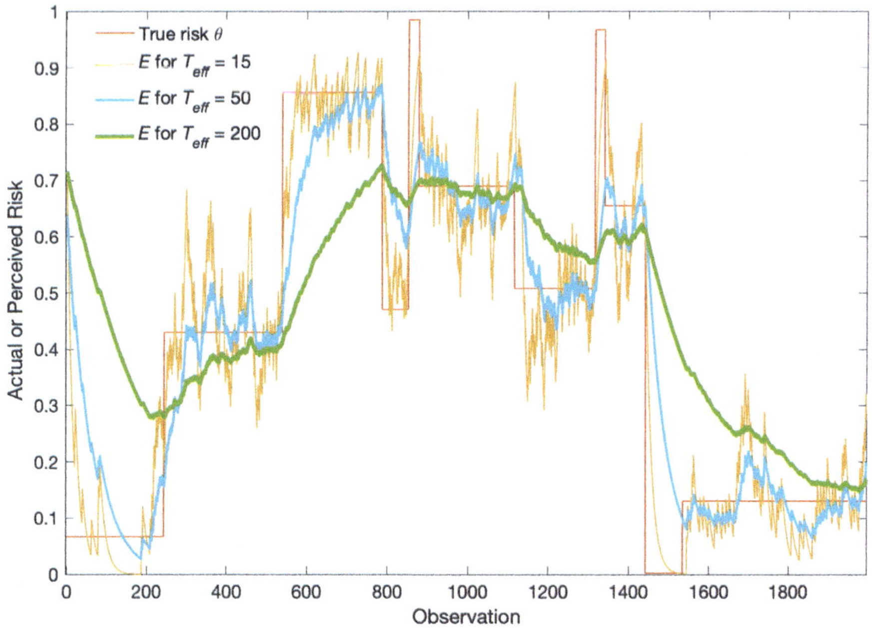

Expected Value of Information

The simplest measure of information loss for report E is $(x-E)^2$, whose mean given the true θ is $\sigma_\theta^2+(E-\theta)^2$. The improvement from a revised report $E+\Delta E$ is $2(\theta-E)\Delta E-(\Delta E)^2$, which for the optimal value $\Delta E=\theta-E$ will shave the information loss to σ_θ^2. When θ is unknown, the best choice is $\Delta E=\langle\theta\rangle-E$, which reduces the mean information loss by $(\Delta E)^2$. If ΔE depends on forthcoming evidence, the expected value of sample information (EVSI) equals $\langle(\Delta E)^2\rangle$.

Agents can be incentivized by offering EVSI as expected reward. The strictly proper scoring rule $S(\Delta E,x|E)=2(x-E)\Delta E-(\Delta E)^2$ offers an expected reward $R(\Delta E)=(\Delta E)^2$ for reporting $\Delta E=\langle\theta\rangle-E$. This is valued at EVSI before the evidence arrives.

For pure Bayesian updates, ΔE after the next observation is $\eta(x-E)$ with marginal expected value $\eta^2(x-E)^2$. The expected reward just prior to observation is $\eta^2\sigma_E^2$, which matches the result (2.5) for $R''=2$. Let's now adjust this for tiny doubts. Multiplying both sides of (3.6) by θ and taking expectations,

$$\begin{aligned}\Delta E&=(1-\lambda)\frac{V}{\sigma_E^2}(x-E)+\lambda\left(\frac{\alpha}{\alpha+\beta}-E\right)\\&\equiv(1-\lambda)\eta(x-E)+\lambda E_{shift}\quad.\end{aligned}\tag{3.7}$$

Imagined shifts add an extra term $\delta=\lambda(E_{shift}-\eta(x-E))$. When θ stays the same, the bias is δ. When θ changes, bias jumps to $(1-\lambda)(E_{shift}-\eta(x-E))$. Weighting the two cases by their probabilities, the EVSI is approximately

$$\eta^2\sigma_E^2+\lambda^2E_{shift}^2\,. \tag{3.8}$$

The second term is usually much smaller than the first. The few exceptions occur where E nears 0 or 1.

With a beta(α,β) prior and $\lambda=0$, the EVSI of x_T is roughly reciprocal to $(\alpha+\beta+T)^2$. This makes early information far more valuable than the rest, even though draws are presumed independent. For $\alpha+\beta=2$, x_1 averages 16 times as valuable as x_{10} and over 50 times as valuable as x_{20}.

While the contrasts are less extreme for higher $\alpha+\beta$, x_1 at $\alpha+\beta=10$ is over 7 times as valuable as x_{20}.

To maintain beta priors with $\lambda>0$, let's imagine that observers are notified whenever θ shifts but not told the new value. Figure 3.8 plots duration on a log scale for another simulation of $(\alpha,\beta,\lambda)=(1,1,0.005)$, along with T_{eff} for a Perfect Learner who isn't informed of the shifts. T_{eff} is generally much lower. This points to extra skittishness and jitter except at the moments θ changes. Observers informed of true duration concentrate overwhelmingly on the early observations after a shift. Perfect Learners try to emulate but never know for sure which x are early or late.

Suppose we record EVSI for ten million plays and order the values from highest to lowest. Figure 3.9 displays cumulative shares of total EVSI with and without notification of shifts in θ. When timing is known, the first two observations after a shift—1% of the total—provide half the total EVSI. The most influential 10% account for 94% of total value, the most influential 20% account for nearly 98%, and the bottom 50% only 0.3%.

Figure 3.8: T_{eff} with and without notification of shifts in θ

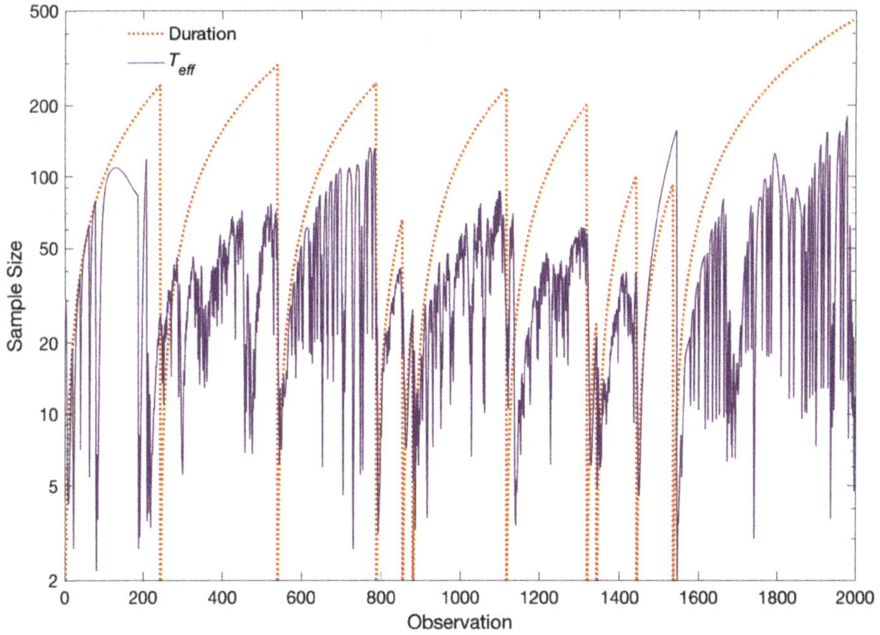

Figure 3.9: Cumulative shares of aggregate EVSI

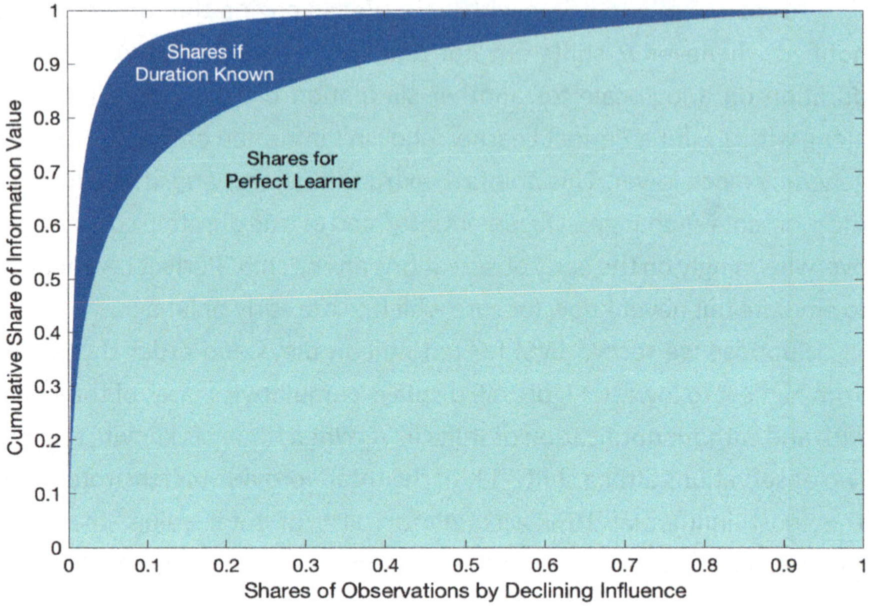

Information value is more dispersed for a Perfect Learner but still highly concentrated. The top 1% of observations account for 29% of total EVSI, the top 10% account for 70%, and the top 20% account for 81%.

Compared to EMAs and other rolling averages, Tiny Doubters look very skittish. They seem to care far too much about a small portion of information. Yet they would be even pickier if they knew when θ changed.

Higher $\alpha+\beta$ dilutes concentration as the variance after reset shrinks. Higher λ dilutes concentration as faster reset prunes observations that provide little marginal information. For frequent small changes in risk, there's not much point to being picky. Tiny Doubts help most when scanning for rare major shifts.[6]

4

Cumulant Hierarchy

Mandelbrot (1963) distinguished between calm and chaotic market phases. He noted that small changes in market prices tend to be followed by small changes and big changes by big changes. The observations helped inspire his discovery of fractals. Later Mandelbrot (1997, also with Hudson 2007) likened market price movements to physical turbulence and used fractals to illustrate their common connection. However, the resemblance might be superficial. Where does it come from?

The prime source is a feature that rational learning shares with the flows of physical fluids. Called a "cumulant hierarchy", it induces cascades that can either amplify or smooth random fluctuations. Since cumulant hierarchies aren't easy to grasp and typically lack neat solutions, let me make them more palatable by sketching the context of their discovery.

Modern physics traces both smooth and turbulent flows to a single differential equation known as Navier-Stokes (Navier 1822, Stokes 1851). It equates the force on each portion of fluid to mass times acceleration. Conservation of momentum ties flow nonlinearly to changes in flow. A key driver is the ratio of momentum to internal friction (viscosity), now known as the Reynolds number (Reynolds 1883). Low Reynolds numbers are associated with smooth flows. High Reynolds numbers trigger turbulence.

Despite Navier-Stokes' one-line simplicity, the strange behavior it induces is hard to figure out. The mathematical physicist Horace Lamb allegedly joked:

> When I die and go to heaven, there are two matters on which I hope for enlightenment. One is quantum electrodynamics and the other is the turbulent motion of fluids. And about the former I am rather optimistic. (Lamb 1932)

The renowned physicist Feynman, who illuminated the perplexities of quantum electrodynamics, called turbulence the most important unsolved problem of classical physics:

> [T]here is a physical problem that is common to many fields, that is very old, and that has not been solved.... Nobody in physics has really been able to analyze it mathematically satisfactorily in spite of its importance to the sister sciences. It is the analysis of circulating or turbulent fluids. (Feynman 1963, Chapter 3)

No one has managed to model turbulent flow in a way that fails to diverge over space and time. Even supercomputers, which update Navier-Stokes over massively detailed grids, can't yet predict major storm paths reliably for more than a few days at a time. Mathematically, each moment depends on the moment one order higher. It is an endless ladder. According to one textbook,

> Closing the moment hierarchy [...] is the underlying problem of turbulence theory. (McComb 1990, p.7)

The effect is clearest when we shift to cumulants, which are moments adjusted for the impact of lower moments. The first cumulant κ_1 is the mean E, the second cumulant κ_2 is the variance V, the third cumulant κ_3 equals skewness times $V^{3/2}$, and the fourth cumulant κ_4 equals (excess) kurtosis times V^2. When the higher cumulants dissipate, flows are smooth. When higher cumulants surge, they create turbulence.

Imagine two nearly adjacent drops of fluid. As the surrounding forces are very similar, only very high cumulants will register any noticeable differences. Nevertheless, each cumulant drives changes in the cumulant below, which given sufficient stress can make the two drops behave very differently. That is how turbulence arises. Reduce stress enough and the higher cumulants shrink toward zero. That is how turbulence subsides.

Rational learning is also governed by a single equation and it too generates a cumulant hierarchy. In the case of Brownian motion, we will see that all neighboring cumulants satisfy

$$\text{volatility}(\kappa_n) = \frac{|\kappa_{n+1}|}{\sigma}. \tag{4.1}$$

The more Gaussian the beliefs, the smoother the associated learning, as Gaussian is the only distribution with all higher-order cumulants equal to zero (Marcinkiewicz 1938). In practice, no beliefs are perfectly Gaussian, so any outliers that boost higher cumulants will tend to resonate below and induce turbulence.

Judging from (4.1) alone, high turbulence can easily spiral out of control. However, the core learning question also includes strong mean-reverting elements for every cumulant beyond the first. Notably, V shrinks at rate V^2/σ^2 absent apprehensions of change in underlying risk. This is bound to dominate once E gets close to the true mean θ of x and explains why turbulence subsides. But precise tracking of the surges and retreats in turbulence is bound to be difficult.

Hansen (2007) acknowledged the practical challenge in updating the distribution of beliefs, which "for many state spaces can be an infinite dimensional state variable" and in principle should be updated every infinitesimal instant. However, he declared this "a computational issue, not a conceptual one". Rewriting the optimal updating rule in terms of cumulants shows that Hansen was mistaken. Updating a never-ending cumulant hierarchy is indeed a conceptual issue, since it explains why beliefs converge in some contexts and diverge in others.

Since most capital market prices are closely related to mean beliefs E, the volatility V/σ of E boosts price volatility beyond that generated by dividends and other economic fundamentals. The uneven volatility of V in turn makes price volatility unevenly volatile. Thus (4.1) with $n=1$ is the core cause of the excess volatility documented by Shiller (1989, 2000), while (4.1) with $n=2$ is the core cause of the GARCH behavior first documented by Engle (1982). One equation explains two Nobel Prize-winning discoveries without invoking foolishness.

Derivations

The rest of this chapter elaborates these points, first by deriving the salient equations and then by simulating their impact on equity prices. As noted in Chapter 2, Brownian motion $dx = \mu dt + \sigma dz$ can be viewed as the limit of a binary process with outcomes $\pm\sigma\sqrt{dt}$ and mean μdt. This implies the Bayesian update $dp(\mu) = (\mu - E)(dx - Edt)/\sigma^2$, which can be rewritten

$$dp(\mu) = p(\mu)S(\mu)dy \quad \text{for } S(\mu) \equiv \frac{\mu - \kappa_1}{\sigma} \text{ and } dy \equiv \frac{dx - \kappa_1 dt}{\sigma}. \qquad (4.2)^7$$

$S(\mu)$ indicates the idiosyncrasy of the belief in standard deviations. In financial terms it is the Sharpe ratio (Sharpe 1966) of a bet on dx when the price is κ_1 and the true drift is μ. The noisy term dy, which measures the unexpected news in standard deviations, has unit volatility. Hence (4.2) can be interpreted as

% change in conviction = idiosyncrasy × surprise .

Note that p is a function of μ alone; σ is considered known. This implicitly presumes that any discrete observation on Δx over a unit time interval can be subdivided into N discrete observations $\Delta x_1, \ldots, \Delta x_N$ at times $1/N$ apart. While this does nothing to estimate a constant drift better, the sum of $(\Delta x_i)^2$ estimates σ^2 much better than $(\Delta x)^2$ does. The bias drops to μ^2/N from μ^2 and the variance drops to $2\sigma^2/N$ from σ^2. In principle we can identify σ as quickly and precisely as needed.[8]

Let us now fold in tiny doubts. Let $\lambda_{\mu\omega}$ denote the instantaneous Markov switching rate from current value μ to another value ω. Weighting outflows and inflows by perceived likelihoods, the full update works out to

$$dp(\mu) = p(\mu)\frac{\mu - \kappa_1}{\sigma}dy + \left(\int p(\omega)\lambda_{\omega\mu}d\omega - p(\mu)\int \lambda_{\mu\omega}d\omega\right)dt. \qquad (4.3)$$

This was first discovered by Wonham (1964). For a rigorous proof, see Liptser and Shirayev (1977).

Multiplying both sides of (4.2) by $e^{s\mu}$, integrating, and denoting the moment-generating function by M,

$$dM(s) = \frac{\langle \mu e^{s\mu} \rangle - \kappa_1 \langle e^{s\mu} \rangle}{\sigma} dy = \frac{M'(s) - \kappa_1 M(s)}{\sigma} dy . \qquad (4.4)$$

We then multiply $M + dM$ by another moment generating function M^{shift} to capture the Markov shifts.

The cumulant generating function K is the logarithm of M. Applying Itô's lemma, it has volatility $\chi \equiv (K' - \kappa_1)/\sigma$ and drift $K^{shift} - \frac{1}{2}\chi^2$, so

$$dK(s) = \left(K^{shift}(s) - \frac{(K'(s) - \kappa_1)^2}{2\sigma^2} \right) dt + \frac{K'(s) - \kappa_1}{\sigma} dy \qquad (4.5)$$[9]

Hence dK has a shift component, a mean-reverting component and a volatile component, with all three changing over time. The mean reversion tends to straighten K and shrink higher cumulants. However, the higher cumulants in the volatility potentially allow even a tiny K^{shift} to stir up considerable turbulence. Equating powers of s in the Taylor expansions,

$$d\kappa_n = \left(\kappa_n^{shift} - \frac{1}{2\sigma^2} \sum_{j=1}^{n-1} \binom{n}{j} \kappa_{j+1} \kappa_{n-j+1} \right) dt + \frac{\kappa_{n+1}}{\sigma} dy , \qquad (4.6)$$

where κ_n^{shift} denotes the anticipated Markov shift in κ_n. This holds for every whole number n and confirms (4.1).

Only one kind of Markov shift rules out turbulence. That's where μ changes through Brownian motion with known drift and volatility. Then all higher-order $\kappa_n = 0$ and V is never volatile. Extensions to multiple dimensions are known as Kalman filters (Kalman 1960) and are widely used in navigational tracking.[10]

I will call (4.5) the Core Learning equation and include all broadly equivalent formulations under this rubric. I have repeatedly invoked its $n = 2$ corollary in (4.6). We now see an endless chain of implications that might be expressed as "the fastest learner has highest V", "the most flexible learner has highest κ_3", and so on. Neher and Shraiman (2011) derived a similar hierarchy for biological evolution and used it to refine Fisher's fundamental theorem of natural selection.

To apply the Core Learning equation to equity pricing, let Brownian x describe log dividends. Let $\bar{Q}(\mu)$ denote the fair price-to-dividend ratio given a known current drift μ, which makes $Q \equiv \langle \bar{Q}(\mu) \rangle$ the fair ratio given beliefs p. The combination of (4.3) and a first-order Taylor series expansion of $\bar{Q}(\mu)$ around $\bar{Q}(E)$ implies

$$dQ_{bayes} = \int \bar{Q}(\mu) dp(\mu) d\mu = \left(\int \bar{Q}(\mu) \frac{\mu - E}{\sigma} p(\mu) d\mu \right) dy \quad \Rightarrow$$
$$dQ = \frac{\text{cov}(\bar{Q}, \mu)}{\sigma} dy + \left\langle Q^{shift} \right\rangle \approx \bar{Q}'(E) \frac{V}{\sigma} dy + \left\langle Q \text{ migration rate} \right\rangle dt. \tag{4.7}$$

Suppose $\bar{Q}' > 0$, as is true whenever higher current μ bodes well for future μ. Then dQ is positively correlated with dx and creates excess volatility.[11] Since the volatility ς of fair market price $P = Qe^x$ is defined in log terms, we can express this more precisely as

$$\varsigma = \sigma + \text{vol}(\log Q) \approx \sigma + \frac{\bar{Q}'(E)}{Q} \cdot \frac{V}{\sigma} = \sigma \left(1 + \frac{\bar{Q}'(E)}{Q} \eta \right). \tag{4.8}$$

The relative excess volatility is the product of two terms: the relative sensitivity $\bar{Q}'(E)/Q$ of Q to E and the learning rate $\eta = V/\sigma^2$, which indicates the sensitivity of E to news. Applying (4.6) with $n = 2$,

$$dV = \left(\kappa_2^{shift} - \frac{V^2}{\sigma^2} \right) dt + \frac{\kappa_3}{\sigma} dy. \tag{4.9}$$

This supports the modeling of ς as volatile. Among GARCH models, the popular exponential variant (Nelson 1991) approximates our specification best. However, since GARCH presumes a determinate response to dy, (4.9) predicts that stochastic volatility or SV models (Taylor 1986, Heston 1993) potentially fit equity data better. Ghysels et al (1996), Andersen and Benzoni (2009), and Christoffersen et al (2010) confirm superior SV fits.

Excess volatility is closely associated with the value of information. When aiming to reduce the mean squared error in E using a tiny observation dx, the second term in (3.8) shrinks to insignificance. This makes the EVSI proportional to $\eta^2 \sigma^2 = V^2/\sigma^2$. For decisions involving an equity valued at Qe^x, the relevant information costs hinge on the relative errors in $Q-\bar{Q}(\mu^*)$. Applying (4.7) and (4.8), the relevant learning rate applied to $dx-Edt$ is

$$\eta_Q \equiv \frac{\left\langle \bar{Q}(\mu^*)S(\mu)\right\rangle}{\left\langle \bar{Q}(\mu^*)\right\rangle \sigma} \approx \frac{\bar{Q}'(E)}{Q}\eta , \tag{4.10}$$

which matches the relative excess volatility in (4.8). The corresponding EVSI is $\eta_Q^2\sigma^2$.

For application, consider reset at rate λ to a distribution π. If all impatience and risk aversion are folded into a constant discount rate r, we can calculate fair Q as follows. Given drift μ of log dividends, an investment of $\bar{Q}(\mu)$ returns dividends at rate 1 and grows at expected percentage rate $\mu+\frac{1}{2}\sigma^2$. Let $\langle\bar{Q}\rangle_\pi$ denoted the expected post-reset value of Q. Applying the discount factor $e^{-rdt}\approx 1-rdt$, equilibrium requires

$$\bar{Q}(\mu)\approx 1\cdot dt+(1-rdt)\left(1+\mu dt+\tfrac{1}{2}\sigma^2 dt\right)\left((1-\lambda dt)\bar{Q}(\mu)+\lambda dt\langle\bar{Q}\rangle_\pi\right). \tag{4.11}$$

Taking the limit, $(\lambda+J(\mu))\bar{Q}(\mu)=1+\lambda\langle\bar{Q}\rangle_\pi$, where $J(\mu)\equiv r-\frac{1}{2}\sigma^2-\mu$ denotes the dividend yield when $\lambda=0$. Hence

$$\bar{Q}(\mu)=\frac{q(\mu)}{1-\lambda\langle q\rangle_\pi} \quad \text{for} \quad q(\mu)\equiv\frac{1}{\lambda+J(\mu)}; \tag{4.12}$$

$$Q=\left\langle\bar{Q}(\mu)\right\rangle\approx\bar{Q}(E)+\tfrac{1}{2}\bar{Q}''(E)V=\bar{Q}(E)\left(1+q^2(E)V\right). \tag{4.13}$$

A negative value implies infinite valuation. To keep valuations finite, we must have $\max(\mu)<\lambda+r-\frac{1}{2}\sigma^2$ and $\lambda\langle q\rangle_\pi<1$.

For two drifts with high μ_H, low μ_L, $\pi_H\equiv\pi(\mu_H)\equiv 1-\pi_L$ and $p\equiv p(\mu_H)$, (4.3) simplifies to

$$dp=p(1-p)S_{MAX}dy+\lambda(\pi_H-p)dt , \tag{4.14}$$

where $S_{MAX}\equiv(\mu_H-\mu_L)/\sigma$ denotes the maximum possible Sharpe ratio. This implies a learning rate $\eta=p(1-p)S_{MAX}^2$ and relative excess volatility

$$\frac{\varsigma}{\sigma}-1=\frac{\eta}{\lambda+J_H+p(\mu_H-\mu_L)}. \tag{4.15}$$

It follows that ς varies a lot if S_{MAX} is high and λ is low.

The long-term (stationary) density h^* of p can be computed from the Kolmogorov forward equation as

$$h^*(p) \propto \left(p^{2\pi_H-1-U} (1-p)^{1-2\pi_H-U} \exp\left(-\frac{\pi_H}{p} - \frac{1-\pi_H}{1-p} \right) \right)^{2/U}, \quad (4.16)^{12}$$

where $U \equiv S_{MAX}^2 / \lambda$ and distills four free parameters into one. U indicates how clearly the two states can be distinguished. (For the Kelly gamblers described in the next chapter, U equals the maximum expected profit per unit of capital given perfect information about the current drift.) Local peaks occur where

$$p\left(2Up^2 - 3Up + U + 1\right) = \pi_H. \quad (4.17)$$

When U is small, h^* is single-peaked. As U gets larger, h^* develops two peaks, with the peak for the more common drift higher and closer to the edge. Figure 4.1 displays a few examples.

Figure 4.1: Stationary densities for various U and π_H

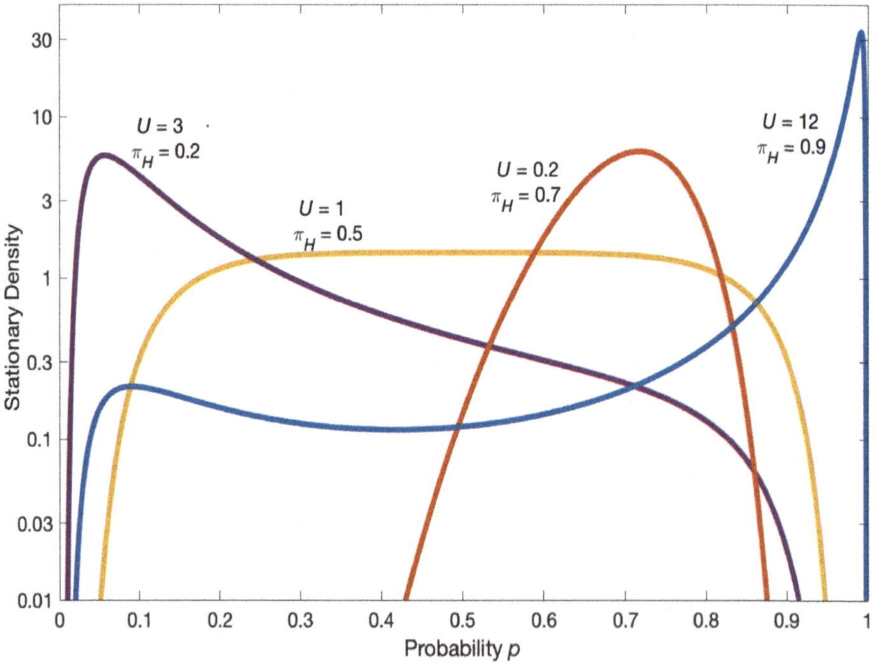

Bear markets with low E are generally rarer than bull markets with high E. Hence, rational learning predicts higher average ς in bear markets than bull markets. Intuitively, high μ is the norm, so its longer duration gives time to build confidence that μ is high. This associates it with p close to 1 and diminished $p(1-p)$, η, and $\varsigma-\sigma$. Low drift, given less time to build confidence that drift is low, is associated with less extreme p and higher $p(1-p)$, η, and $\varsigma-\sigma$.

This is easiest to prove when U is large and π_H is close to 1, since that implies a big cluster of bullish beliefs very close to 1. More generally, note that κ_3 in a two-state world is proportional to $(1-2p)V$. When bull markets are the norm ($\kappa_3<0$), disappointing news ($dy<0$) will boost V via (4.9), which in turn boosts ς via (4.8).

Simulations

How big are these effects likely to be? For insight, let's track Q and ς in models fit to empirical data on disaster risks gathered by Barro and Ursúa (2012). Defining disaster as a cumulative decline in real per capita consumption or GDP of at least 10%, they estimated average durations of 25 years in normal times and 3.64 years in disasters. They also estimated an average growth rate of 1.88% with a standard deviation of 6.24% and a mean drawdown during disaster of 21.1%.

To fit a two-drift Markov switching model, a first pass solves $(\lambda\pi_L,\lambda\pi_H)\approx(1/25,1/3.64)$ to estimate $\lambda\approx0.315$ and $\pi_H\approx0.872$. Divide the mean log drawdown by mean duration to estimate $\mu_L\approx-0.0652$. Set $\sigma=0.0624$ as reported. Then reduce the average percentage growth rate by $\frac{1}{2}\sigma^2$ to estimate $\langle\mu\rangle\approx0.0169$, which in turn implies $\mu_H\approx0.0289$.

However, the first pass presumes that observed log GDP growth matches its drift, when in fact random noise might trigger spurious switches. Restriction to yearly data further coarsens the comparisons. In a detailed search using Monte Carlo simulations and geared to minimizing the average absolute log deviation from the empirical statistics, the best-fitting mix was $(\sigma,\mu_H,\mu_L,\lambda,\pi_H)=(0.054,0.0238,-0.118,0.288,0.950)$.[13]

Barro and Ursúa modeled an aggregate equity holding that pays out aggregate consumption and priced It as if every payout is iid with the average risk of disaster. In their model, agents never notice any near-term persistence in dividend trends. Agents never change their views on a fair price-to-dividend ratio. As a result, price volatility ς always matches dividend volatility σ. No real-life market behaves that way.

Suppose we modify this to allow traders to learn from experience. We can plug the best-fit parameters and a discount rate r into (4.3), (4.12), and (4.13) to simulate market pricing. Figure 4.2 charts Q and ς for a 200-year simulation at $r = 0.05$ with updates 250 trading days per year.

In a million-year simulation, Q ranged from 23 to 34 with mean 33.7. ς ranged from 5% to 30% with mean 7.3% and root mean square (square root of mean variance) 8.3%. Hence, price volatility as usually measured averages about 50% higher than dividend volatility σ.

The most striking feature of Figure 4.2 is the inverse relation between Q and ς. The correlation is -0.71. U is nearly 24, which from (4.17) implies an upper peak density at $p \approx 0.9998$ and a lower peak at $p \approx 0.04$. Mean p

Figure 4.2: Q and ς for 200-year simulation of two-drift model

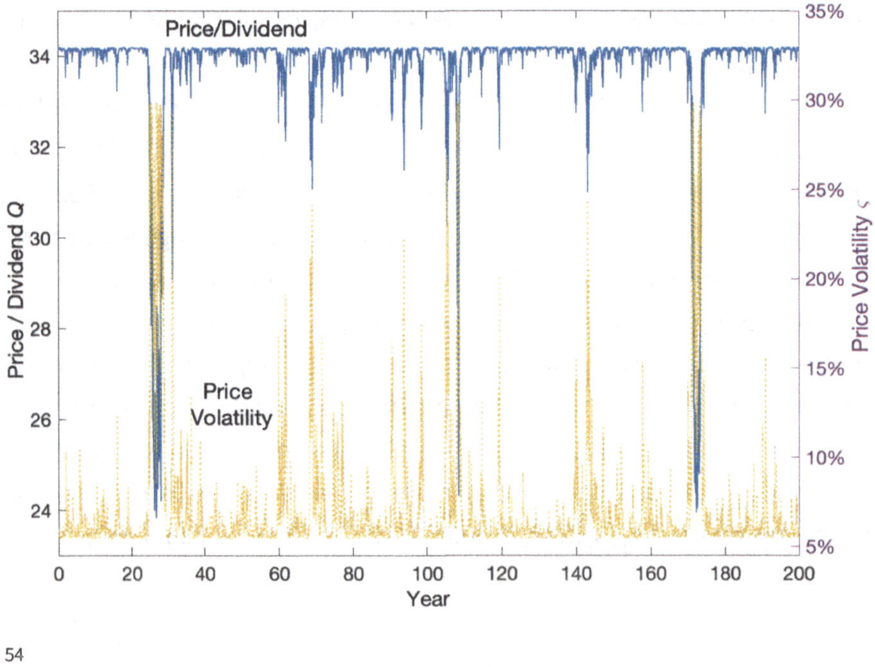

is 0.978 when drift is high and 0.415 when drift is low. As a result, ς averages 1000 bps higher for low drift than for high drift.

Despite the wide range of ς in Figure 4.2, the market usually looks very calm, often for years at a time, due to extremely high confidence that drift is high. For more realism, we can replace the fixed disaster shock with variable shocks. Imagine that drift equals a constant μ_{MAX} less a random gamma-distributed disturbance with shape D and mean μ_{GAM} that resets at rate λ. A shape $D \ll 1$ makes most outcomes cluster near zero while stretching a dispersed minority well above the mean. See Figure 4.3.

To fit the data, σ and λ should not change much from the two-drift model while μ_{MAX}, μ_{GAM}, and D should lie near μ_H, $\pi_L(\mu_H - \mu_L)$, and π_L respectively. The best $(\sigma, \lambda, \mu_{MAX}, \mu_{GAM}, D)$ combination is approximately $(0.052, 0.315, 0.0296, 0.0129, 0.12)$. Half of the gamma outcomes are less than 0.0002, while 20% exceed μ_{GAM}. A severe crisis with mean contraction of 10% or more per year starts on average once in 160 years.

Figure 4.4 displays a 200-year simulation of Q and ς using the best-fitting gamma-reset parameters, with beliefs updated daily over a grid of

Figure 4.3: Gamma distributions for various shapes

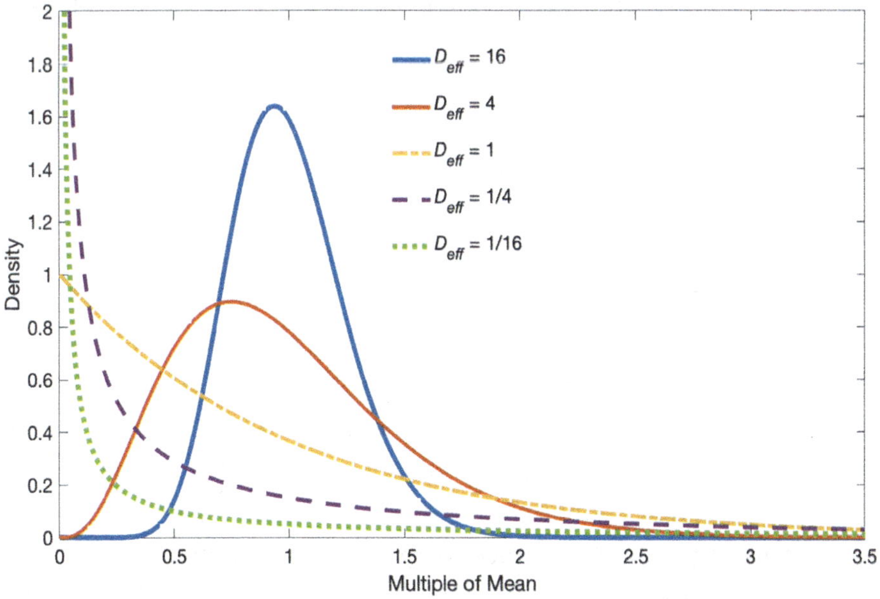

Figure 4.4: Q and ς for 200-year simulation of best-fit gamma-reset model

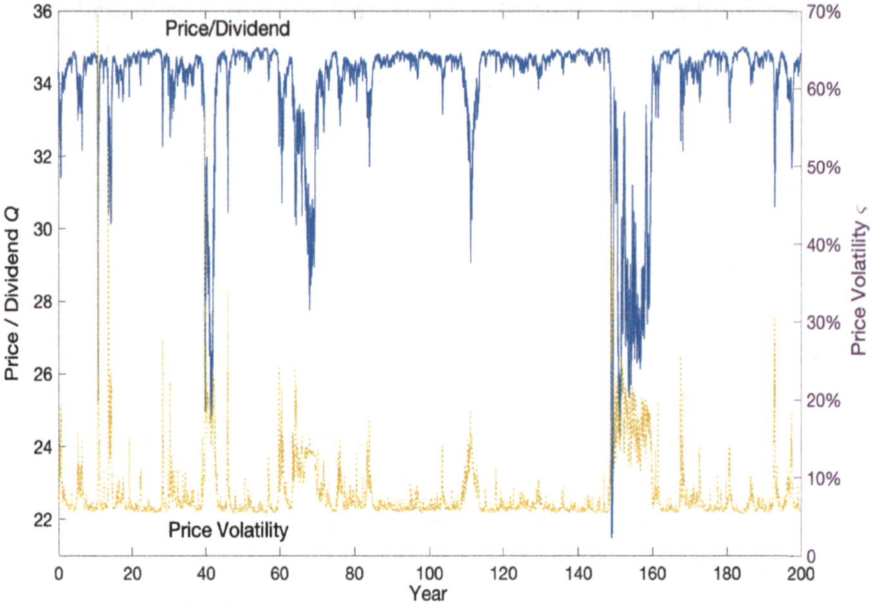

300 possible μ. Both Q and ς vary much more than in Figure 4.2. In most other respects the two models are similar. The gamma-reset means of 34.0 for Q, 7.7% for ς, and 9.0% for root mean square ς are close to the two-drift values.

The variations in marginal informativeness are huge. In our gamma reset model, 1% of observations provide 54% of total EVSI, 5% provide 80% of EVSI and 10% provide 90%. Nearly all this clustering stems from variation in V as its correlation with η_Q is 0.99. The median η is 0.094 with associated duration of 10.6 years. For the 10% least certain beliefs, the median η is 0.76 with duration 1.3 years.

The most disconcerting regularity is Q's tendency to hug a fixed ceiling whenever volatility recedes. Q lies within 1% of its maximum 85% of the time with two drifts and 61% of the time with gamma reset. This is best addressed by allowing for uncertainty about the true parameters.

For illustration, suppose an Almost-Perfect Learner (APL) knows that a two-drift or gamma-reset process prevails with intermediate $\sigma = 0.053$ and has narrowed down to three possible values $\{b_i - \delta_i, b_i, b_i + \delta_i\}$ for every other parameter. The b_i denote the best fit. The δ_i equal 0.005 for μ_{GAM}, 0.01 for π_H, μ_L, μ_L, or max(μ), 0.05 for D, and 0.1 for λ. Suppose the APL allows a once-per-century chance of reset from a uniform distribution and updates beliefs 250 trading days per year.

Figure 4.5 displays Q and ς for a 200-year simulation, where the actual drift, like in Figure 4.3, is gamma-reset with best-fit parameters. Here Q fluctuates far more in normal times with higher ς. In a 100,000-year simulation, Q ranged from 9 to 57 with mean 36.8 and standard deviation 5.6. ς ranged from 6% to 472% with mean 10.1% and root mean square 10.9%.[14]

The extra uncertainty makes the price charts look much more realistic and much less rational. The average Q in one century can differ markedly from another, as if long-term optimism alternates with long-term malaise. Yet the long-term risk-generating process never changes and the market never fails to assess the evidence dispassionately.

Figure 4.5: Q and ς for 200-year simulation of gamma reset with Almost-Perfect Learner

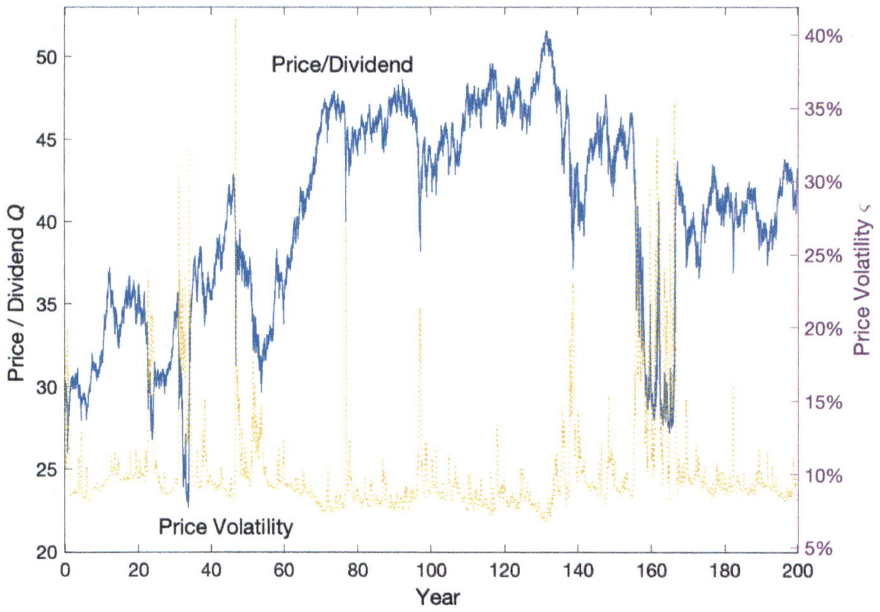

There is abundant evidence on the peculiar behavior of equity volatility. Let's check how well or badly our APL model conforms to it, drawing on 25 million daily observations over 100,000 simulated years.

Excess volatility: Price volatility ς averages 2.0 times the dividend volatility σ, without any irrational exuberance or any other squandering of useful information. This matches the 2.0 ratio in data maintained by Shiller (2025) on yearly real dividends and returns for S&P 500 since 1871. While the ratio for monthly Shiller data is even higher at 2.9, the extra differential can mostly be explained by the short-term smoothing of dividend payments.

Heteroskedasticity: The standard deviation of ς is about 40% of its mean. The best empirical comparator we have is the VIX index, which infers the predicted S&P 500 volatility embedded in option prices (along with a risk premium) and has been recorded since 1986. VIX too has a standard deviation of about 40% of its long-term mean.

Scaling of volatility shocks: ς is about 0.5 correlated with absolute daily changes $|\Delta\varsigma|$. Empirically, the correlation between VIX and $|\Delta\text{VIX}|$ is slightly lower at 0.37.

Higher bear market volatility: ς is about -0.5 correlated with drift μ. ς averages 575 basis points (bps, where 1 bp = 0.01%) higher in years with net losses exceeding 10% than in years with net gains. ς rises an average 360 bps more if the following year registers net losses too. Empirically, Maheu and McCurdy (2000) found that volatility tended to rise in bear markets and to keep rising with bear market duration.

Asymmetric volatility: $\Delta(\log\varsigma)$ and $\Delta(\log P)$ are negatively correlated, with values swelling from -0.18 daily to -0.54 yearly. The correlations between returns and next-period absolute returns are weaker but still negative, ranging from -0.01 for daily returns to -0.18 for yearly returns.

The negative correlations are well documented empirically (Beakert and Wu 2000, Talpsepp and Rieger 2010, Horpestad et al 2019). They were first noted by Black (1976), who attributed them to firm equity becoming riskier as its value falls relative to debt. This view gained so much credence that asymmetric volatility is often referred to the "leverage effect". However, Hasanhodzic and Lo (2019) found similar correlations in firms free of debt. Our focus on higher uncertainty during downturns offers a more robust explanation.

Volatility clustering: Absolute daily returns $|\Delta(\log P)|$ are 0.19 correlated with $|\Delta(\log P)|$ 5 days previously. Taylor (2005, Table 4.10) reported the corresponding empirical correlations of 0.21 for the S&P 500, 0.26 for the Nikkei 225 and 0.14 for the FTSE 100.

Long memory: ς is remarkably persistent. Figure 4.6 charts the autocorrelation functions (ACF) out to 250-day lags for ς and $|\Delta(\log P)|$. The autocorrelation for ς is 0.99 at one day and 0.80 at one month. If memory were purely short-term, the ACF would decay exponentially to 0.5 for one quarter and 0.07 for one year. Instead, the ACF is 0.61 for one quarter and 0.38 for one year. The ACF for absolute daily returns behaves similarly; it is always close to 20% of the ACF for volatility.

Figure 4.6: Autocorrelations for ς and absolute daily returns

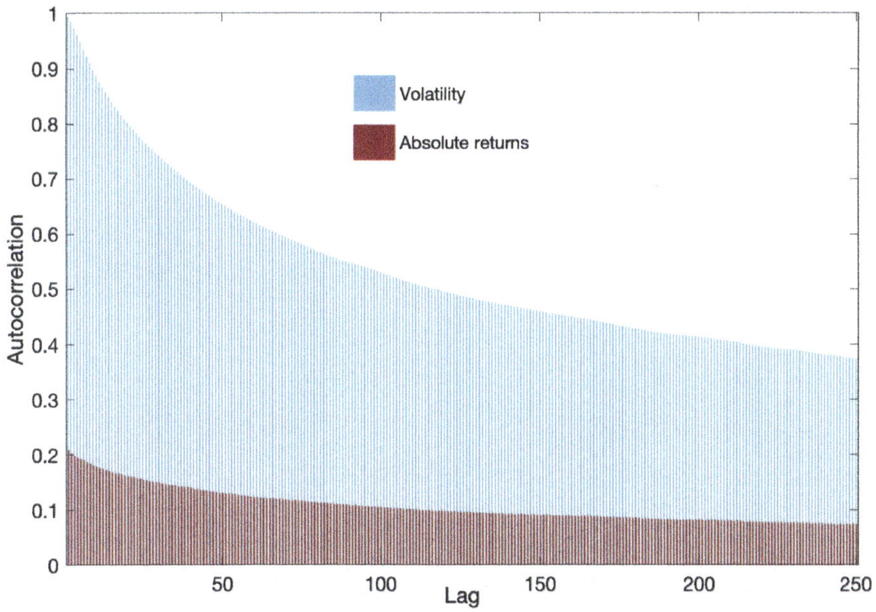

The best fit is a 38:62 mixture of two exponential decays, one at rate 0.029 and the other at rate 0.002. Extending the ACF out to lags of 2000 rejects hyperbolic decay, where the ACF is a power function of the lag and the sum of correlations is unbounded. However, the 0.05 correlation at 8 years' lag affirms the high persistence.

Empirically, absolute returns have long been known to be highly persistent. Goodhart and O'Hara (1997) called it "perhaps the most intriguing feature revealed by empirical work on volatility". According to Ding et al (1993), autocorrelations nearly halve after 100 days, which holds for our model too, and their model fit for the ACF implied like ours a finite bound for correlations. Andersen and Bollerslev (1997) favored hyperbolic decay, as have several other researchers. However, Cont (2001) emphasized the slow convergence of sample ACFs for long lags and this is readily confirmed in our model. ACF behavior often varies significantly from one century-long sample to another.

In summary, our model of near-perfect learning about disaster risks accounts very well for observed equity volatility. It generates all the main features noted by researchers. The implied price paths behave broadly like real-world prices. No other models in the literature can approach this degree of verisimilitude without imposing ad hoc constraints.

5

Prediction Markets

In FIRE models, the various agents form identical probability assessments. These are known as homogeneous beliefs. Everyone wants to be long at prices higher than the common mean E and short at prices less than E. The only price that can clear markets is E, which inclines agents to stick with what they have. When beliefs change, E changes, but since every agent considers the new E fair that does not induce trading.

Trading requires heterogeneous beliefs, some of which likely stray far from Perfect Learning. Heterogeneous beliefs in turn raise questions about how a consensus market price forms and evolves. For example, the previous two chapters assumed Bayesian updates with small probability λ of reset from a long-term distribution π. In real life, there isn't the slightest chance that everyone will agree on either λ or π. How then is fair market price determined? How does it evolve?

Our main result is that the Core Learning equation applies nearly as well to a competitive market as it does to an individual. Even though traders compete vigorously against each other and share no information other than the bids that determine price, the market largely behaves like a unified rational learner who weights beliefs in proportion to traders' effective capital. I call this the Invisible Mind theorem.

The Invisible Mind theorem is the Bayesian counterpart to theorems in economics that liken ideal market allocation to ideal central planning. It sets Hayek's (1945) praise for market pooling of knowledge on a solid analytic foundation. The invisible mind that forecasts value complements the reputed invisible hand that steers production.

What makes this work is a gambling strategy intended to maximize long-term growth of capital. It is known as the Kelly criterion (Kelly 1956) and colloquially as Fortune's Formula (Thorp 1960, 2000, Poundstone 2005). It advises gamblers to size independent bets in direct proportions to capital and expected profit per bet and in inverse proportion to the conditional variance per bet. In ideal conditions, the best proportionality coefficient on the combination is one:

$$\text{Optimal Size of Bet} = \text{Capital} \times \frac{\text{Expected Profit}}{\text{Conditional Variance}} \tag{5.1}$$

Suppose a security paying x is tentatively offered at price F with conditional variance σ_F^2. Suppose Kelly gambler i, whose share of total gambling capital K is $k_i \equiv K_i/K$, is confident the risk is θ_i. Applying (5.1), gambler i aims to own $K_i(\theta_i - F)/\sigma_F^2$ securities, where a negative value indicates a short position. To make this feasible, let's assume securities can be traded in any fractional amount. For supply to match demand, price must equal the capital-weighted mean belief. Since the net payoff to each security is $x - F$, the outcome changes each capital share by

$$\Delta k_i = k_i(\theta_i - F)\frac{x-F}{\sigma_F^2} \quad \text{where } F = \sum_i k_i\theta_i \,. \tag{5.2}$$

If we replace every k_i with subjective probability $p(\theta_i)$, (5.2) matches the Bayesian update (3.4) for binary risk. If we replace every θ_i with $\mu_i dt$, F with $\langle \mu_i \rangle dt$ and σ_F^2 with $\sigma^2 dt$, (5.2) matches the Bayesian update (4.2) for Brownian risk. As for the component reflecting tiny doubts, we can replicate that through wealth taxes coupled with lump-sum transfers or with each gambler i treated as a large pool of individuals who defect probabilistically to other pools. While the replication will never be exact, the differences seem no more significant than any other departure from pure Markov switching. The Kelly market behaves like a Perfect Learner with each k_i matched to p_i.

To justify (5.1), consider a population of gamblers who plow all their winnings back into new bets and have no other source of funds. Since the gross return over multiple periods is the product of the gross returns at each period, gamblers with the highest log growth rate of wealth will eventually dominate all others. Kelly strategies aim to deliver that by maximizing the expected log growth rate of wealth.

For Bernoulli risk θ_i, a gambler expects $w_i = \omega_i K_i$ securities costing F each to generate expected log growth

$$\theta_i \cdot \log(1-\omega_i F + \omega_i) + (1-\theta_i)\log(1-\omega_i F). \tag{5.3}$$

This is maximized when $\omega_i = (\theta_i - F)/\sigma_F^2$ or $w_i = K_i(\theta_i - F)/\sigma_F^2$. If every gambler behaves similarly, $F = \sum_i k_i \theta_i$ and (5.2) holds. Similarly, given Brownian volatility σ and drift μ_i, a gambler holding $w_i = \omega_i K_i$ claims costing $F\,dt$ each expects log growth

$$\begin{aligned} \langle \log(1+\omega_i(dx - F\,dt))\rangle &\approx \langle \omega_i(dx - F\,dt) - \tfrac{1}{2}(\omega_i dx)^2 \rangle \\ &= \omega_i(\mu_i - F)dt - \tfrac{1}{2}\omega_i^2\sigma^2 dt \;. \end{aligned} \tag{5.4}$$

This is maximized when $w_i = K_i(\mu_i - F)/\sigma^2$, which again matches (5.1).

While it is less widely recognized, Kelly principles can also be applied to categorical decisions. Picking the best categorization can be likened to gamblers picking the winner in a horse race.

Let each horse $n \in \{1,\ldots,N\}$ have an associated bet that pays 1 if n wins and 0 otherwise, with no take for the track. Since a single horse wins, the market prices $\{F_n\}$ sum to 1. This makes a portfolio of M bets on each horse equivalent to placing no bets at all. By adding this portfolio and adjusting M, we can assure that the initial cost of bets for any given strategy is zero.

Let θ_{ni} denote gambler i's view on fair price for a bet on horse n and let B_{ni} denote the amount bet. Kelly principles advise to

$$\text{maximize} \sum_{n=1}^{N} \theta_{ni} \log(K_i + B_{ni}) \quad \text{subject to} \quad \sum_{n=1}^{N} F_n B_{ni} = 0\,. \tag{5.5}$$

First-order conditions require $\theta_{ni}/(K_i + B_{ni}) = \gamma_i F_n$, where γ_i denotes the Lagrange multiplier on the constraint. Rearranging and summing over n,

$$1=\sum_{n=1}^{N}\theta_{ni}=\gamma_i K_i \sum_{n=1}^{N}F_n+\sum_{n=1}^{N}F_n B_{ni}=\gamma_i K_i\cdot 1+0=\gamma_i K_i\,.$$

Hence $\gamma_i=1/K_i$ and $B_{ni}=K_i(\theta_{ni}-F_n)/F_n$. While this looks different from (5.1), it implies the same change in capital share k_i when n wins and matches the corresponding Bayesian update:

$$\Delta k_i=k_i\frac{\theta_{ni}-F_n}{F_n}\,. \tag{5.6}$$

How should gambler uncertainty about θ or μ change the previous equations? At first glance, it shouldn't. If we take expectations of (5.3)-(5.5) with respect to p, all that seems to matter is the subjective mean E or more precisely the expected profit $E-F$. Most Kelly related literature disregards the uncertainty and so did I in Osband (2020).

However, as Baker and McHale (2013) correctly emphasize, what really matters is the objective risk, which E estimates with noise. The noise calls for additional caution in gambling. For the Brownian case $E_i=\mu_i+\varepsilon_i$, consider a bet of $f(E_i-F)/\sigma^2$, where f is a shrinkage parameter applied to (5.1). Substitution into (5.4) indicates expected profit of approximately

$$f(\mu_i-F)^2\frac{dt}{\sigma^2}-\tfrac{1}{2}f^2(E_i-F)^2\frac{dt}{\sigma^2}\,, \tag{5.7}$$

which is maximized by

$$f=\frac{(\mu_i-F)^2}{(E_i-F)^2}=\frac{(\mu_i-F)^2}{(\mu_i-F)^2+\langle\varepsilon_i^2\rangle}\,. \tag{5.8}$$

Since (5.8) both dampens responses and varies them across gamblers, in principle a speculative market might differ substantially from the Invisible Mind optimum. However, in simulations they evolve similarly. This reflects a combination of

- variance of individual E_i and variance across $\{\mu_i\}$ shrinking when risk seems stable.
- Individual and group variances surging when risk seems unstable.
- lagging adjustment of F inducing more catch-up bets later.
- the practical need to smooth (5.8) since μ_i is not known.

Inter-Threaded Markets

Kelly gambling potentially challenges its own long-term persistence. A Perfect Learner will eventually dominate if there is one, leaving no one to gamble against. However, Perfect Learning is unattainable in practice as there is too much mystery with too little revelation. Real-life markets spontaneously regenerate variance through new risks, new entrants and new ideas. The challenge for theorists is how to model it.

I will model regeneration as reset at constant rate λ to a uniform distribution over traders. I choose constant λ for lack of insight on how it should change and a uniform distribution for lack of insight on which trading strategy can best deal with new surprises. For very high λ, traders maintain similar capital shares and market F is close to the simple average belief. For miniscule λ, a single strategy dominates. That leaves a wide range for λ that makes learning considerably more robust without adding much extra noise.

Instead of trying to optimize λ, let us allow traders to vary not just by current beliefs but also by propensity to migrate. To make this tractable, it is easiest to simulate N different markets, each with a different fixed reset rate λ_n, and then treat these markets as N Kelly traders gambling against each other with aggregate reset rate Λ. I call this combination an inter-threaded market or ITM. Figure 5.1 displays an example with EMAs as base-level traders, an overlay of Kelly markets with reset, and a single market aggregator on top.

Note that the layering does not change the Bayesian updates per se. If we replace the EMAs in Figure 5.1 with Tiny Doubters or add them alongside, the markets linking them change only their shares k_i of credibility capital, not the forecasts of individual EMAs or TDs. If TD_j assigns probability $p_j(\theta_i)$ to risk θ_i, that imparts weight $k_j p_j(\theta_i)$ to the market as a whole. The aggregate weight on θ_i is $\sum_j k_j p_j(\theta_i)$. While the layering restricts the types of regeneration, it allows vastly more freedom than the conventional approach which confines all switching probabilities to zero.

Figure 5.1: Inter-threaded market with base-level EMAs

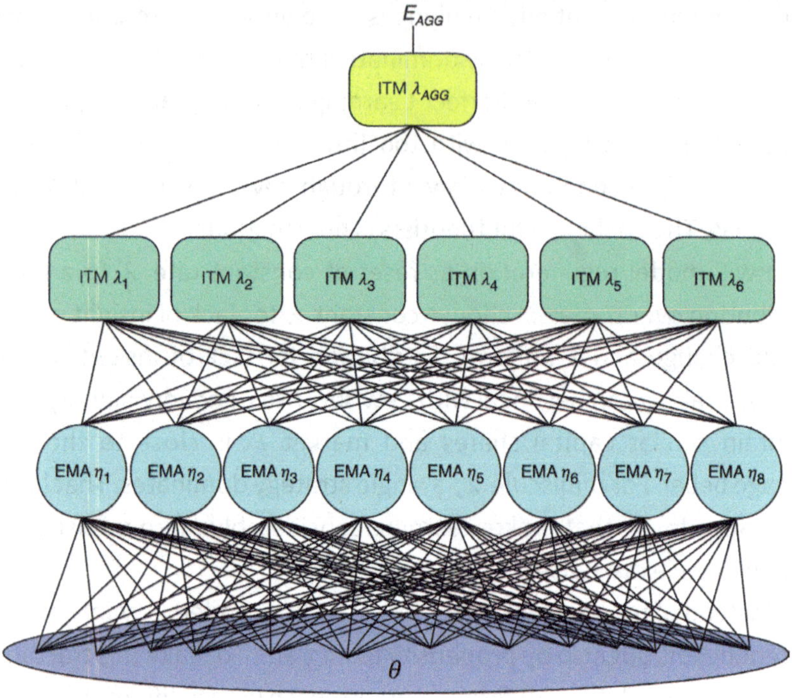

To better appreciate a Kelly-driven market, let's compare it with three fixed-credibility alternatives. Figure 5.2 depicts E for a coin that switches to $\theta=0.9$ from an initial $\theta=0.5$, with observations numbered after the switch. Here 100 Tiny Doubters view the same sample path, including 20,000 observations before the switch. Each TD(α,α,λ) anticipates reset at rate λ from a beta(α,α) distribution but no TDs agree on both values. Their differences span λ from 10^{-2} to 10^{-11} and α from 1 to 10.

How should we form a consensus predictor? Figure 5.3 compares four methods. The first sticks with the best predictor overall. Since stability paid off before the switch, it is the slowest to adjust. The second method gives each predictor one vote. It selects the median E (Black 1948), which adjusts much faster. The third method selects the average E, which begins its shift even sooner but never moves fast and drags out adjustment too long. The fourth method uses a full Kelly ($f=1$) ITM. While it is by far the most skittish method, it tracks the underlying risk best.

Figure 5.2: Mean beliefs after switch from $\theta = 0.5$ to $\theta = 0.9$ with disagreements over reset process

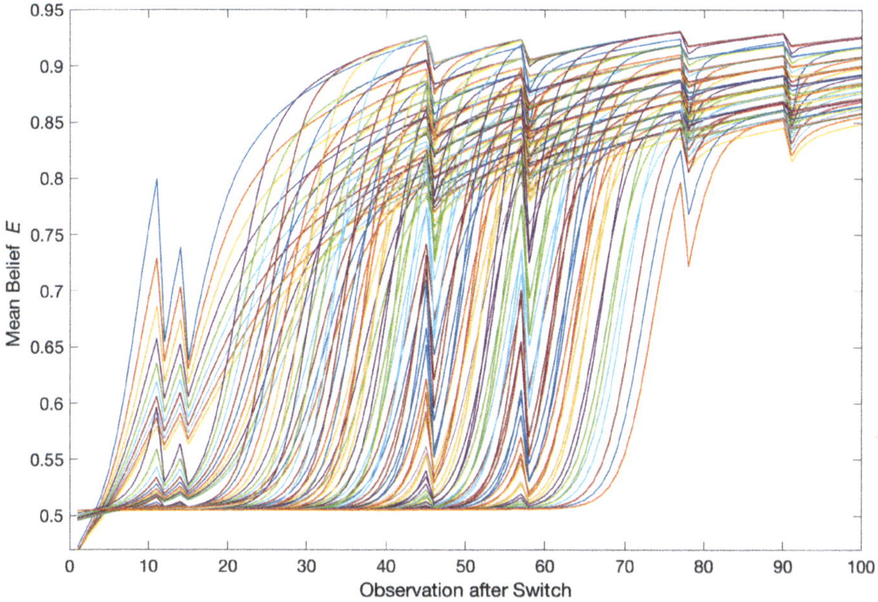

Figure 5.3: Comparison of market, median, mean and best pre-switch predictors for Figure 5.2

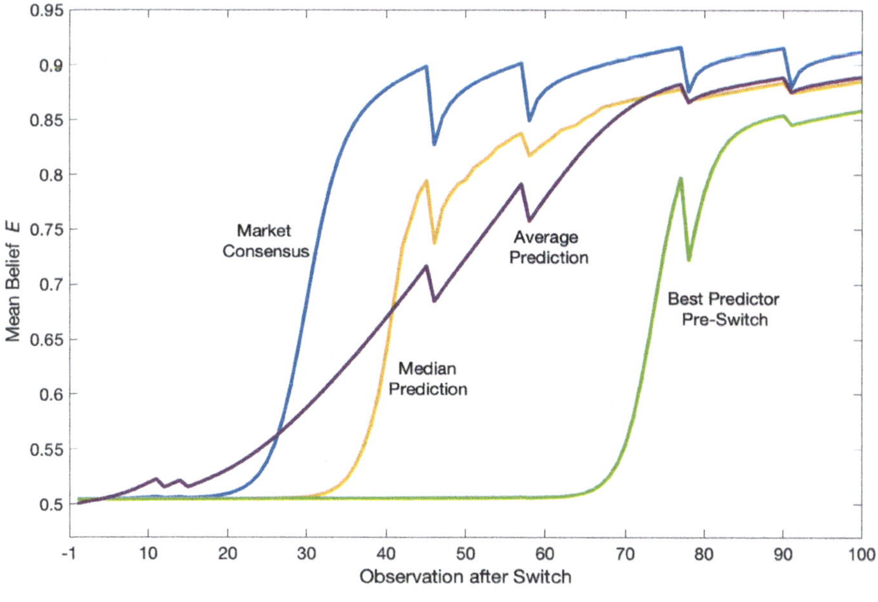

Next let's revisit the Markov example in Chapter 3, where θ is resampled from a uniform distribution at rate 0.005. Our Perfect Learner achieved an RMSE of 0.090 for E versus θ, whereas the best RMSE for an EMA was 20% higher at 0.108. Recalling that the information cost is proportional to the squared error, let us define the relative information loss (RIL) compared to a Perfect Learner as

$$\text{RIL} \equiv \frac{\text{MSE}}{\text{MSE}_{\text{PL}}} - 1 = \left(\frac{\text{RMSE}}{\text{RMSE}_{\text{PL}}}\right)^2 - 1 \tag{5.9}$$

In this case, the best EMA has an RIL of 44%.

Let's see what an ITM of EMAs can accomplish. Consider 16 base-level EMAs with η_i geometrically spaced from 0.0004 to 0.5. In 10 million plays, the mean RIL exceeds 300% and the RIL for the mean of the EMAs is 150%. The base seems weak. Nevertheless, its ITM lowers RIL to 25%.

If we replace EMAs with rolling averages of the same duration, the RIL for the simple mean deteriorates to 240%. Why? Because rolling averages are more vulnerable to older observations' loss of relevance. Yet its ITM achieves an RIL of 34%, not terribly worse than for the EMAs.

In both cases, the ITMs add considerable volatility to the base-level predictors. Figure 5.4 charts learning rates for 2000 observations on an ITM of EMAs, with η_{agg} estimated as $\Delta E_{agg}/(x - E_{agg})$. These generally exceed the η_{PL} for Perfect Learners. This belies the common presumption that calmer investors guarantee calmer markets.

Big jumps are better tracked when Tiny Doubters are included. Consider 36 different base-level TDs with α and β chosen from $\{1,3,9\}$ and λ chosen from $\{10^{-1}, 10^{-2}, 10^{-3}, 10^{-4}\}$. While the RIL for their simple average is 40%, the RIL for their ITM is just 2%. Let me caution that TDs don't always outperform EMAs. If θ is a diffusion with low to moderate volatility, EMAs will track it better. For robustness, include both. ITMs of mixtures of TDs and EMAs will naturally allocate most credibility capital to whichever type predicts best, usually with less than 2% extra RIL.

Figure 5.4: Learning rates for ITM of EMAs

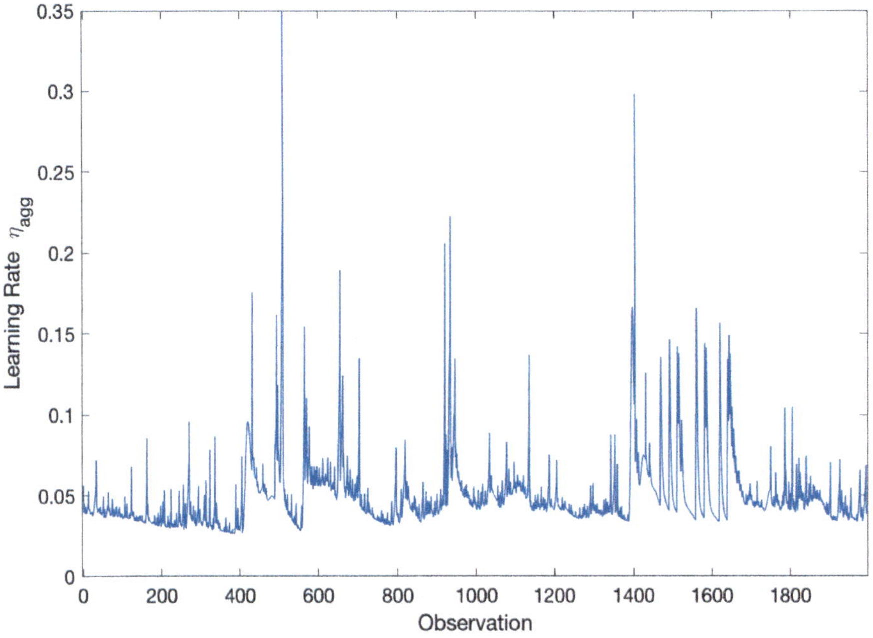

For the dividend drift with gamma reset modeled in Chapter 4, no EMA achieves an RIL less than 110%. An ITM of 16 EMAs with annual η spaced geometrically on [0.01,5] shrinks the RIL to 40%. In contrast, all TDs with parameters within 50% of the benchmark values have RILs of 65% or less and most ITMS of TDs that span those ranges have RILs of less than 2%.

Consider the Almost-Perfect Learner from Chapter 4, who continually updates 162 hypotheses about drift-switching equity dividends. This is basically equivalent to a full Kelly ITM with 162 competing traders. The RIL is 2.5% while $\langle \varsigma \rangle$ is 120 bps higher than for a Perfect Learner.

In general, diverse ITMs predict short-term moves nearly as well as Perfect Learners with only marginally more volatility. Where they stumble is long-term prediction as their grasp of causal mechanisms is poor. For example, a standard EMA always forecasts a constant long-term trend and so do ITMs drawing only on EMAs. While capital markets are great short-term error correctors, they are often sloppy prophets. Chapter 6 will examine this further.

Fractional Kelly

Gambling markets invariably disappoint. Every gambler expects on average to earn a profit, yet the aggregate profit from betting against each other is always zero. This differential persists even if we fold in risk premia to allow profits on positive holdings. From a FIRE perspective, gambling serves no purpose other than to part fools from their money and ceases once capital is consolidated among the wise.

Gambling poses fewer challenges to RATE as each trader expects others to bear the expected loss. However, others' willingness to take the opposite side may give each trader pause. Despite my personal belief in a fair price of E, I might respect the consensus market price F enough to accord it a $1-f$ chance of being correct. This shaves my bet to a fraction f of full Kelly, which if applied consistently is known as fractional Kelly.

Fractional Kelly can be viewed as a type of herding. While the term is often used pejoratively to suggest that people are too dumb or lazy to think for themselves, it is hard to fault some deference to the wisdom of the crowd. In principle, fractional Kelly might be viewed as full Kelly using a revised mean $E^* = F + f(E-F)$.

Fractional Kelly can also stem from risk aversion. Full Kelly gambling is gut-wrenching. Chin and Ingenoso (2007) derived the chances of various drawdowns (maximum percentage losses) given long iid sequences of Kelly-driven bets. For full Kelly, the probability of at least temporarily losing a fraction L of wealth is $1-L$: e.g., 75% chance of a 25% drawdown and 25% chance of a 75% drawdown. Big drawdowns sap convictions and prod losers to bet wildly or throw in the towel.

To reduce those risks, most professional gamblers opt for fractional Kelly, typically with f of one-half to one-quarter. While half Kelly cuts the expected growth rate of wealth by 25%, it trims the probability of a 75% drawdown to less than 2%. While quarter Kelly cuts the expected growth rate by 56%, it trims the probability of a 50% drawdown to less than 1%. The general formulas are $(1-f)^2$ for the reductions in mean log growth rate and $(1-L)^{2/f-1}$ for the drawdown risks (Chin and Ingenoso 2007). Figure 5.5 displays the latter.

Figure 5.5: Drawdown risks under fractional Kelly

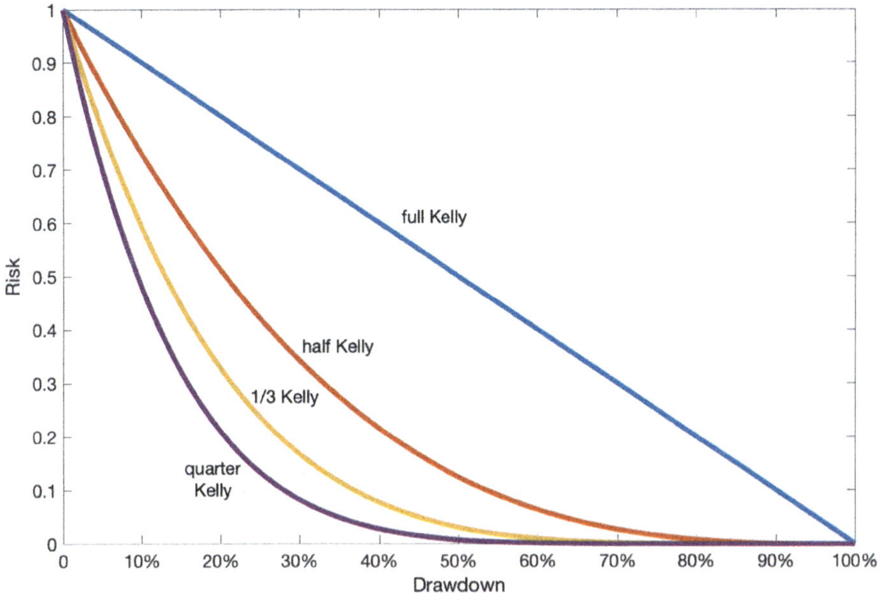

To the extent that people gamble more cautiously than full Kelly advises, it seems unfair to describe them as "irrationally exuberant" or "fooled by randomness". Perhaps they so rue those tendencies that they veer the other way. Perhaps they become too hesitant to bet their own convictions and too fearful of being fooled.

AI inroads in poker support this interpretation. Poker is far more complex than coin flips. Each player sees only part of the evidence and the odds vary with the strategies chosen by others. This places a premium on clever deceit, where humans have long excelled. Nevertheless, in 2017 an AI system "Libratus" routed four top human players in a high-stakes poker tournament, thanks largely to its appetite for risk (Popper 2017):

> One of the things Libratus does well is bluff.... Libratus has also been over-betting frequently, wagering far more to win a hand than is currently up for grabs in the pot... Doug Polk, a poker pro who bested a previous AI [said] "Humans don't really like that. It feels like you're risking a lot of money to win so little. The computer doesn't have that psychology. It just looks at the best play."

How does fractional Kelly affect market behavior? While the impact varies, half to quarter Kelly markets generally behave like modestly smoothed versions of full Kelly markets. Consider again the ITM of 162 competing equity traders. The RILs for half Kelly and quarter Kelly actually improve slightly on full Kelly. Half Kelly prices are 80 bps less volatile than full Kelly prices and quarter Kelly trim another 50 bps. Fractional Kelly pulls consensus toward the average parameter combination, which in this case is the Perfect Learner's.

The similarities have three main causes. First, traders keep adjusting their beliefs regardless of how boldly they bet . Second, sequential updates give cautious traders repeated opportunities to catch up. Third, when risks are misspecified, fractional Kelly helps protect against overbetting.

In principle, a constant Kelly fraction f is inferior to the dynamic adjustment (5.8). However, the latter presumes knowledge of the objective risk θ or μ, which is ruled out by positive V. The best proxy we have is our mean belief E, but when tested in simulations, the proxied (5.8) always performs significantly worse than the best fixed f. Half to quarter Kelly is far more robust.[15]

Sometimes fractional Kelly is essential. Suppose a trader discovers a stream of non-public information X that objectively shifts drift $\mu(t)$ by $\delta(t)$. If the rest of the market is oblivious to X and the trader's capital share is small, the optimal full Kelly bet is $\delta(t)/\sigma^2$. Yet the advantage isn't certain; others might have gained this information too or surrogates for it. So the trader needs to estimate what share of $\delta(t)$ is already incorporated into price and reduce the Kelly share accordingly. The information most eagerly sought on trading floors is Δt advance notice of relevant public news, in which case the optimal bet is $\left(\delta(t+\Delta t)-\delta(t)\right)/(\sigma^2 \Delta t)$.

Hence, we can view a fractional Kelly capital market as a decentralized learning machine. Compared to an ideally unified learner, it performs more redundant calculations and weighs more evidence inappropriately. However, it is highly robust, scales easily through replication and minor tweaks, and saves enormously on horizontal information flows.

This is no mere theoretical possibility. Bill Benter started developing a system in 1984 for probabilistic gambling on horse races. He collected all the relevant data he could find, fit logistic regressions to the data and refined the factors to improve the signal to noise ratio. He sized bets using fractional Kelly, which he learned gambling on blackjack in Las Vegas. According to Benter (1994), his betting system was only moderately profitable until he incorporated the market odds into his forecasts. Benter later called this his single most important innovation (Chellel 2018). The system is reputed to have generated nearly $1 billion in profits. In short, Benter successfully utilized all three features that I flagged above as crucial to decentralized learning: probabilistic forecasting, fractional Kelly bets, and feedback from market consensus.

Fractional Kelly markets with redistribution potentially offer insights for neuroscience as brains are decentralized learners too. While the prefrontal cortex is considered the hub of higher cognition, various subdivisions weigh in on its judgments and so do other parts of the neocortex. Here are some challenges to leading theories that Kelly-with redistribution might resolve:

- Bayesian Brain Theory (Knill and Pouget 2004) does not explain how Bayesian updating can occur without centralization.
- Neuroeconomics (Glimcher 2010) models brains as competitive markets but pays little attention to probabilistic inference.
- Predictive Coding (Friston 2009, 2010) simplifies Bayesian learning to minimization of (free-energy prediction error but does not readily account for volatile changes in the learning rate η.
- Thousand Brains Theory (Hawkins 2021) posits largely independent cortical columns but is vague about the "voting" mechanism that governs their interactions.

Compared to neuroscience, machine learning has paid far more attention to mechanisms for decentralization. Bagging models (Breiman 1996) classically rely on majority vote or average vote but some give extra weight to better-performing models. Bagging methods are also used in Federated Learning (McMahan et al 2017), where individual models use private, unshared data. Boosting models (Freund and Schapire 1997) dynamically reweight both data and predictors.

Stacking models (Wolpert 1992) use the predictions of base models as data. Bayesian Model Averaging (Hoeting et al 1999) updates relative credibility. Bayesian Hierarchical Stacking (Yao et al 2022) extends this to multiple layers. Both Bayesian methods have been used in deep learning (e.g., Li et al 2024, Elsemüller et al 2024). Variants are often incorporated into larger AI models under the names Meta-Learning or Super Learning.

A sophisticated literature (e.g., Chen and Pennock 2010, Chakravorti et al 2023) recommends more explicit incorporation of prediction markets into AI learning. Kelly gambling might promote more flexible and efficient coordination with substantially lower energy costs. However, I have not yet found any discussion of Kelly-type gambling in AI, let alone advocacy.

6

Heterogeneity

The most striking feature of major capital markets is the huge volume of trading. It proves that they are rife with disagreement. Yet both FIRE and standard Bayesian analysis imply this should be rare. When risk is fixed, the accumulating evidence eventually impels near-agreement between learners, each of whom is nearly certain. The main rationale that leaves for trading is management of portfolio risk, which is usually depicted as leverage or dilution of a single aggregate exposure.

Grossman and Stiglitz (1980) attributed trading to the costs of acquiring information. These encourage some agents to become better informed and profit from trading with the less informed. However, this cannot explain the trading surges that occur on the release of earnings or jobs reports. De Long et al (1990) attributed trading to poorly informed "noise traders" who stir up the noise they profit from and tend to lose more in risk-adjusted returns than they gain. However, this begs the question common to all behaviorist critiques of FIRE, namely why doesn't the market eventually wise up?

Harrison and Kreps (1978) attributed trading to competing views of the transition rates of a hidden Markov process. Each trader is wrongly certain in ways that allow all traders to hang on. Again, this begs the question of why traders don't learn from their mistakes.

As we have seen, ITMs can continually regenerate heterogeneity despite the attenuation implicit in Bayesian updating. All they need is occasional changes in risk and some kind of bounded rationality—say, tiny exogenous shifts in beliefs (redistribution) or totally fixed beliefs. This chapter explores those connections and draws three main conclusions. First, market pressures to agree on longer-term structural views are far weaker than pressures to agree about stable short-term risks. Second, calm periods allow a broad variety of structural views to flourish. Third, turbulence intensifies disagreements and stirs bigger shifts in capital.

Let us assume a market of fractional Kelly gamblers, with each applying the same fraction f. This is less restrictive than it appears, since an f_i – Kelly gambler with capital K_i behaves short-term like an f – Kelly gambler with capital $(f_i/f)K_i$. Each gambler i expects a rate of profit on capital of $f(E_i-F)^2/\sigma_F^2$, which equals f times the squared Sharpe ratio S_i^2. If $\langle\cdot\rangle_k$ denotes the capital-weighted mean, then

$$\langle \text{rate of profit} \rangle_k = f\langle S_i^2 \rangle_k . \tag{6.1}$$

Given news $\delta = x - F$, $\Delta E_i \approx \eta_i \delta$ and $\Delta k_i \approx f k_i S_i \sigma_F^{-1} \delta$. The deviations reflect redistributions independent of trading that are usually relatively small. To first-order approximation,

$$\begin{aligned} \Delta F &= \langle E_i + \Delta E_i \rangle_{k+\Delta k} - \langle E_i \rangle_k \approx \langle \Delta E_i \rangle_k + \textstyle\sum_i E_i \Delta k_i \\ &\approx \langle \eta_i \rangle_k \delta + f \langle E_i S_i \rangle_k \sigma_F^{-1} \delta = \langle \eta_i \rangle_k \delta + f \langle S_i^2 \rangle_k \delta . \end{aligned}$$

Hence the aggregate learning rate η_{agg} is approximately gamblers' capital-weighted expected profit plus their capital-weighted learning rate:

$$\eta_{agg} \approx \frac{\Delta F}{\delta} \approx \langle S_i^2 \rangle_k + \langle \eta_i \rangle_k . \tag{6.2}$$

Even when each Kelly gambler is convinced that they know the true risk, some slippage will likely occur through redistributive taxes and conversions. Let's model this as shifts in capital backing from μ_i to μ_m that occur at rate λ_{im} for every i and m, irrespective of gambling. Then the aggregate market behaves like a Tiny Doubter with $\eta_{agg} = \langle S_i^2 \rangle_k$.

Kelly trading volumes will be broadly proportional to mean absolute Sharpe ratios $\langle |S_i| \rangle_k$, which are highly correlated with $\langle S_i^2 \rangle_k$. Equity price volatility ς equals $(1+\eta_{agg})\sigma_F$ from (4.8) and (4.10). Hence (6.2) makes three predictions that command much empirical support (Gallant et al 1992, Andersen 1996, Kouba and Slim 2019):

i. ς has a substantial stochastic component that does not depend on trading volume.
ii. ς responds only weakly to low trading volumes.
iii. ς responds strongly to high trading volumes.

I am not aware of any other explanation that so readily accounts for (i)-(iii). For example, if we treat trading volumes as proxies for information flows—a popular theme in the research literature—we might let relevant trading time and observation noise σ^2 scale with volume. While this can plausibly explain (iii), it cannot explain (i) and is muddled about (ii).

The core prerequisite of trading is diversity of capital. I will measure diversity as the reciprocal of the classic Herfindahl-Hirschman Index (HHI) (Herfindahl 1950, Hirschman 1964), which sums the squares of market shares. With N equally sized competitors, the reciprocal HHI equals N.

The more immediate impetus to trading is disagreement, which I will quantify using (6.1). The pre-trade optimism about expected profits matches the post-trade disappointment since total capital is fixed for gamblers betting against each other. Letting $V_{traders}$ denote the capital-weighted variance of traders' mean beliefs E_i, we have

$$\text{Disagreement} \equiv f\left\langle S_i^2 \right\rangle_k = f\frac{V_{traders}}{\sigma_F^2}. \tag{6.3}$$

Recall Figure 3.2, which charts the near-logistic responses of 125 Tiny Doubters (α,β,λ) to a previously fair coin repeatedly landing heads. Let's embed those TDs in a market that started 10,000 tosses earlier, with full Kelly gambling, initially equal capital shares k_i and no redistribution separate from gambling. Figure 6.1 charts capital diversity and disagreement for 50 consecutive heads on a log scale with $f=1$. As disagreements of less than 1 bp are noisy in log terms, I averaged values across 100 simulations.

After 10,000 fair tosses, disagreement about the next risk averages less than 0.1 bp. Yet this tolerates a wide variety of structural views. Capital

Figure 6.1: Capital diversity and disagreement before last 50 heads with $f = 1$

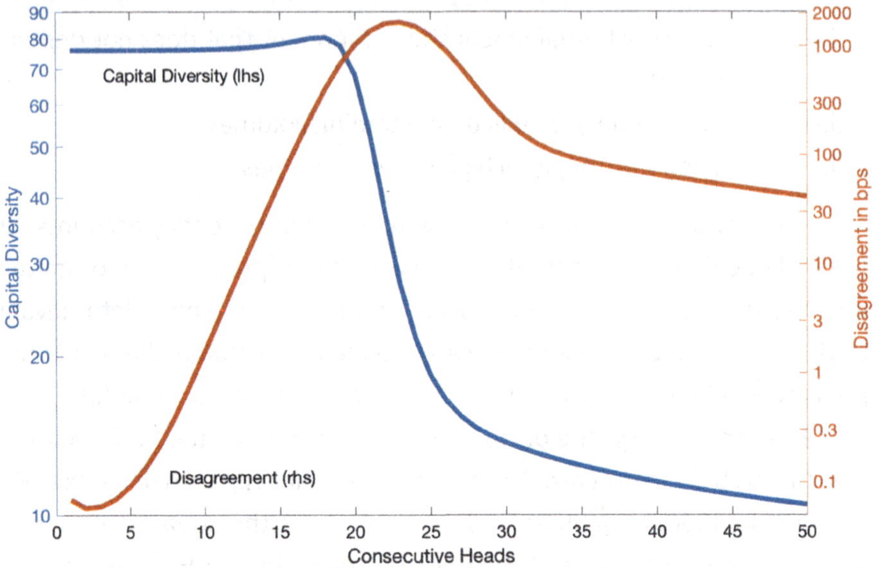

diversity averages 76. Diversity initially edges up as heads repeat but plummets below 20 once early adjusters pocket big winnings. In contrast, disagreement surges near-exponentially to 1600 bps after 22 straight heads and then retreats by 90% after 8 more heads. After 50 straight heads, disagreement shrinks to 40 bps, still 500 times the average pre-shock level.

Trading volume and impact depend far more on disagreement than on capital diversity. Since longs are matched by shorts, the weighted longs equal half the weighted absolute deviations, which for a normal distribution average to $\sqrt{2/\pi} \approx 0.80$ times the standard deviation. Fat tails reduce the multiplier but rarely by more than 20%.[16] Hence

$$\text{Total Longs} = \frac{fK}{2\sigma_F^2}\sum_i k_i \left|E_i - F\right| \approx f\frac{(0.36 \pm 0.04)\sqrt{V_{traders}}}{\sigma_F^2}K. \tag{6.4}$$

In this example, longs exceed 10% of K in value after 18 straight heads, peak at 36% of K at the 24th head, and still exceed 14% at the 50th head. Trading transfers over 5% of K for heads 18 through 24 and peak at 11% of capital for the 21st head. Correlation with disagreement is 0.87.

For an extended game, let traders bet on a Bernoulli risk θ that repeatedly resets with probability $\Lambda = 0.01$ from a uniform distribution. Again I posit 125 Tiny Doubters but with a tighter distribution than before. Each holds to a distinct $(\alpha, \beta, \lambda = 10^{-i})$ for $\alpha, \beta, i \in \{1, 1.5, 2, 2.5, 3\}$ with tiny chance $\lambda_{agg} \leq 0.01$ each round of reset from a uniform distribution. For each (λ_{agg}, f) pair, let's run 100 simulations with 30,000 tosses each.

When $\lambda_{agg} = 0$, the Perfect Learner $(1, 1, \Lambda)$ eventually squeezes out all other traders but only after many thousands of observations. While bad learners lose capital relatively quickly, it Is hard to identify the best learners. The median time for capital diversity to fall below 2 is about 5,000 with full Kelly, 12,000 with half Kelly, and 23,000 with quarter Kelly.

When $\lambda_{agg} = 10^{-3}$, mean capital diversity trends downward for the first 5,000 observations and stabilizes thereafter at 18 for full Kelly and 82 for quarter Kelly. However, in any given simulation the capital diversity continues to fluctuate with standard deviation exceeding 15.

How much does this heterogeneity worsen market forecasts? Remarkably little. Quarter Kelly with $\lambda_{agg} = 10^{-3}$ has a 2% higher RMSE than a Perfect Learner (4% RIL) and generates 12% less price volatility. To observers ignorant of the true drivers, the Perfect Learner might seem significantly less rational than its proxies.

Trading fades gradually when $\lambda_{agg} = 0$. With full Kelly, the winning payoffs for long positions average about 6% of capital at toss 100, 4% at toss 1000, 2% at toss 2500 and 1% at toss 5000. Quarter Kelly reduces the initial share but slows its decline; by toss 8000 it trades more than full Kelly.

In contrast, $\lambda_{agg} = 10^{-3}$ quickly stabilizes average total longs at much higher levels: 5.2% of capital for full Kelly and 2.2% for quarter Kelly. The average market value is half of that. However, there is enormous volatility, with changes averaging about 20% of the means and occasionally surging much higher. Figure 6.2 charts a sample for 1000 observations. It also displays disagreement, which is 0.82 correlated with total longs.

The variance V_{agg} of beliefs that drives price volatility equals the sum of $V_{traders}$ and the capital-weighted variance $\langle V_i \rangle_k$ of traders' $\{E_i\}$. Hence both higher disagreement and higher individual uncertainty tend to induce greater absolute price changes $|\Delta F|$. Sometimes these two factors

Figure 6.2: Bernoulli disagreement and total longs for $(\lambda_{agg}, f) = (0.001, 1)$

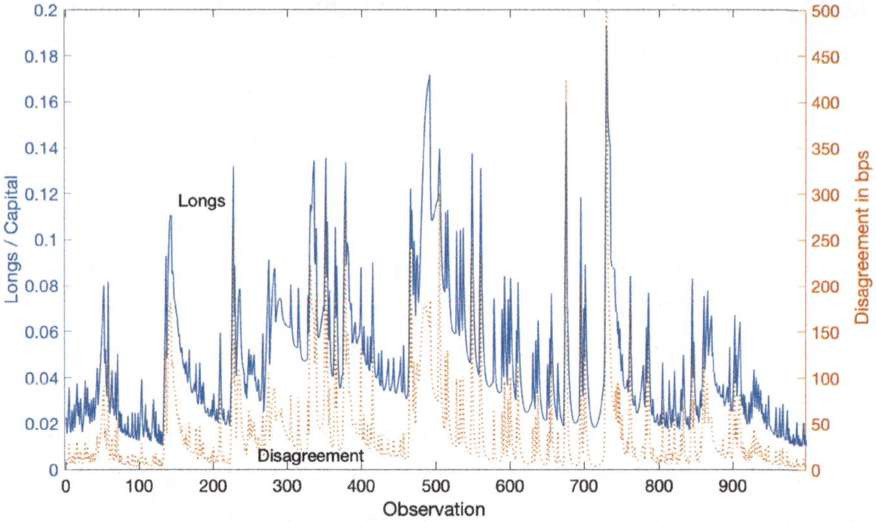

move in tandem, sometimes they don't. When $\lambda_{agg} = 10^{-3}$, $|\Delta F|$ is 0.57 correlated with disagreement for both quarter Kelly and full Kelly; it is 0.30 to 0.36 correlated with trading volume.

Next let's imagine a market that bets continuously on current equity dividends only. While unrealistic, this will help link the previous results to the full equity markets we will study next. Our benchmark will be the ITM of 162 traders from Chapter 5 that replicated the Almost Perfect Learner of Chapter 4. Its yearly $\lambda_{agg} = 0.01$ is 0.4 bps per trading day.

With full Kelly, the mean capital diversity is 99 with a standard deviation of 25 and only 0.1% chance of falling below 15. With quarter Kelly the mean capital diversity is 149 with only a 5% chance of falling below 130 and an 0.1% chance of falling below 60. This testifies to the huge challenge of identifying long-term risks. Even if $\lambda_{agg} = 0$ it takes roughly 3000 years to drive capital diversity below 4.

With full Kelly, disagreement averages 174 bps but is highly uneven. The bottom 90% of values average 131 bps while the top 10% average 562 bps. Quarter Kelly generates average disagreement of 70 bps with similar unevenness. Figure 6.3 charts a 200-year sample on a log scale.

Figure 6.3: Sample dividend disagreement rates for full and quarter Kelly

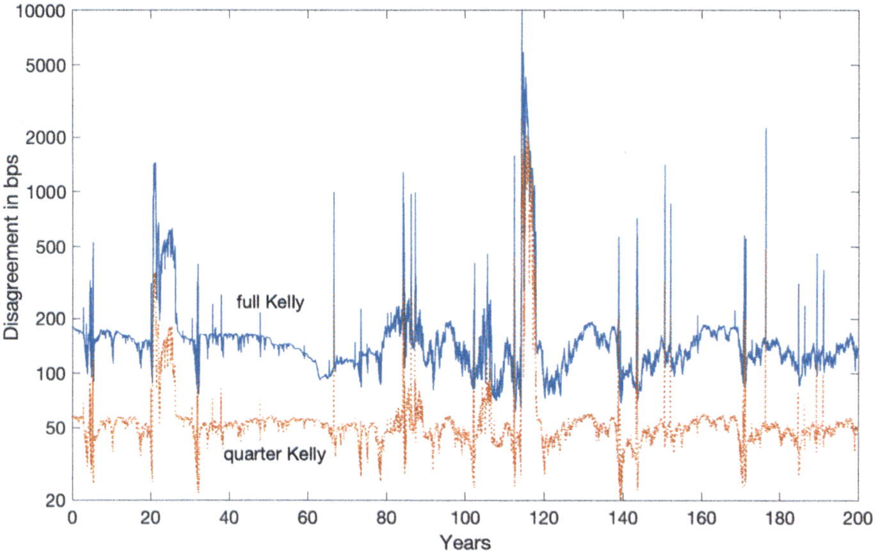

Trading volumes are over 95% correlated with the square root of disagreement so they can be inferred from Figure 6.3. Mean trading volumes are about 60% lower for quarter Kelly than for full Kelly. Standard deviations are about one-third of the mean.

These results are easily extended to a full equity market. From (4.8), equity prices $P = Qe^{x}$ are ς/σ times as volatile as dividends and nearly the same ratio applies to expected returns. To first approximation, the disagreement stays the same and only the position sizes change.[17] Target holdings for each gambler are

$$K_i f \frac{E_i - F}{\varsigma \sigma P}. \tag{6.5}$$

Target trading equals the change in (6.5). The sum of absolute changes multiplied by $P/(2K)$ indicates the total value of trades as share of capital.

Trade value is extremely volatile; the daily log changes have standard deviations exceeding 1.1 for full Kelly and 0.9 for half and quarter Kelly. Figure 6.4 charts a two-year sample with half Kelly.

Figure 6.4: Daily trades with full Kelly for 2-year sample

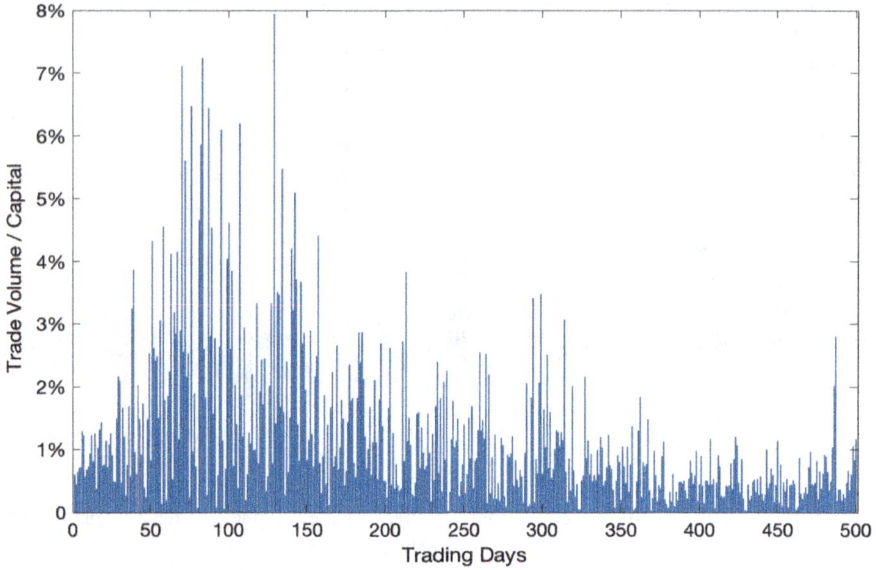

Yearly trading as a share of total capital averages 3.1 for full Kelly, 2.1 for half Kelly, and 1.2 for quarter Kelly. In each case there is a 40% or higher chance of failing to reach even half the long-term yearly average

Trading tends to be greatest when ς is highest. The correlations are 0.34-0.38 for daily values and 0.48-0.55 for monthly averages. Total longs are slightly more correlated with ς and less correlated with trading.

The empirical literature on trading volume, as first summarized by Karpoff (1987), has long noted a strong positive correlation between trading volume and absolute returns. In our simulations the correlation is 0.67. Another empirical regularity is the long memory of trading volumes (Bollerslev and Jubinski 1999). Our simulations exhibit long memory too, with autocorrelations of 0.45 at one week, 0.34 at one month, 0.20 at 6 months, 0.16 at one year, 0.1 at two years and 0.05 at four years.

One more regularity in our simulations is a negative correlation between trading volume and the sign of returns. There is no empirical consensus about the sign of the correlation and multiple researchers have reported that it is positive (Karpoff 1987). This likely reflects practical difficulties in shorting equities, which our model ignores.

Futures Markets

While trading is far more correlated with disagreement than with capital diversity, the core driver is always diversity. In a FIRE world, everyone agrees on core risks; competitive pressures extinguish incorrect views and capital diversity vanishes. In evolutionary terms, only the single fittest view survives. It makes trading very hard to explain.

RATE takes an evolutionary perspective too but like Darwin has more respect for variation.[18] Prospects of change continually breed tiny doubts, which we can view as mutations in structural beliefs. Many have little impact in normal times. They are analogous to nearly neutral mutations of molecules, which are now recognized as crucial to biological evolution (Kimura 1983, Ohta 1992). It takes changing conditions to bring their strengths or weaknesses to the fore.

Finance theorists often make little distinction between near-term and long-term forecasts since both are probabilistic and correctible. However, the core responses to short-term evidence are vastly different. A forecast of mild winter weather is reliably modified when a blizzard moves in. A forecast of long-term global warming is little moved by current weather, perhaps not even if winter is cold worldwide.

Diverse views can thrive in financial markets as long as they predict the short-term well. Conversely, superior long-term perspectives may offer few short-term advantages. Our previous ITM typically maintained capital diversity of over 100 even though one gambler was a Perfect Learner.

Here is a starker example of long-term uncertainty. Suppose a Bernoulli risk θ resets at rate $\Lambda = 0.001$ to a uniform distribution, while a host of "Beta Reset" learners labeled $\text{BR}(\alpha,\beta,\lambda)$ assume reset with probability λ to a $\text{beta}(\alpha,\beta)$ distribution with mean $\mu = \alpha/(\alpha+\beta)$. Let's compare a claim on the next x to a claim on x 10,000 periods from now.

RMSEs for E are relatively close for most BR. The RMSE for the Perfect Learner is 0.045. Divide its λ by 10 or triple its α and the RMSE increases by 8%. Even for $\text{BR}(5,0.5,10^{-5})$, which generally expects a very stable high θ, the RMSE is 0.061, barely one-third above the minimum.

The true future mean $\langle \theta^{fut} \rangle$ is very close to 0.5, as the chance that current θ persists without reset is less than 0.005%. Yet the RMSE for $E^{fut} = \mu + (E-\mu)e^{-10000\lambda}$ is highly sensitive to the specification. It is 0.0002 for a Perfect Learner, 0.107 for BR$(1,1,\Lambda/10)$ and 0.262—little better than a random guess—for BR$(1,1,\Lambda/100)$. For $\lambda > 5\Lambda$, $E^{fut} \approx \mu$, which implies RMSEs of 0.167 for $(\alpha,\beta)=(2,1)$ and 0.25 for $(\alpha,\beta)=(3,1)$.

To improve forecasts, I set up an ITM with 512 different BR predictors on the base layer and three layers of full Kelly TDs above. The base layer offers 8 choices for each parameter, with geometric spacings for $(\alpha,\beta)\in[0.2,0.5]$ and $\lambda\in[10^{-1},10^{-5}]$. The first two higher layers have doubts of 10^{-i} for $i=1,\ldots,8$. The top layer applies the median doubt. I simulate θ, x, $\{E_i\}$, bets and $\{k_i\}$ for 100,000 plays. The future claim is priced as the capital-weighted average of each BR E_k^{fut}.

Figure 6.5 displays RMSEs for E^{fut} averaged across 200 simulations and all observations within 100 periods of the specified time t. It takes about 40,000 observations or 40 switches in θ before E^{fut} tracks the true

Figure 6.5: RMSEs for BR market E and E^{fut} averaged over 200 simulations and all observations within 100 periods of t

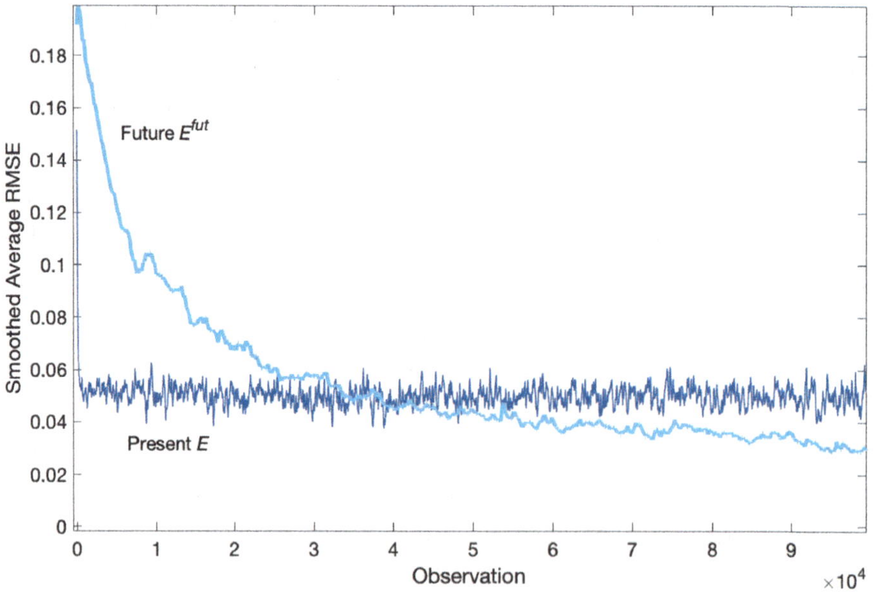

$\langle\theta^{fut}\rangle$ better than E does. Even after 100,000 observations or 100 switches in θ, E^{fut} often diverges from the nearly constant $\langle\theta^{fut}\rangle$ by 0.02 or more.

Suppose we try to improve E^{fut} by focusing on 10,000-day outcomes. In 100,000 days, we obtain only 10 independent observations. Augmenting them with 90,000 overlapping 10,000-day observations doesn't improve the RMSEs much as the overlaps are too highly correlated. Even if the roughly 100 different θ are measured perfectly, they won't reveal the true mean. For random draws from a uniform distribution, the standard error of E is 0.029. That is not much better than the BR market achieves.

Adding to the problems, Bayesian updates that rotate among bad long-term predictors don't always generate good long-term predictors. For example, suppose we replace the base-level BRs with "EMA Reverters", which nudge E toward the current x and then let it revert toward an assumed long-term mean μ. I let μ, EMA duration, and reversion speed take eight values each, with the first spaced linearly on the unit interval, the second spaced geometrically between 0.3 and 0.003, and the third spaced geometrically between 0.001 and 0.0001.

While the consensus E fits θ relatively well, E^{fut} fits 0.5 worse than the average of 0.5 and E. Alternatively, we might estimate E^{fut} as the simple average of all observed x. Figure 6.6 plots the smoothed RMSE for 200 simulations. The trajectory mostly parallels that of the BR market. Eventually the simple average will so outperform E^{fut} for EMA Reverters that it is bound to spur a more BR-type modeling approach.

How would an active market in E^{fut} affect this? Changes in E^{fut}, like changes in equity prices, often hinge on news about current x. Hence contracts in E^{fut} tend to behave like rescaled contracts in E with simpler transfer and settlement. The immediate inferences from x do not change. In principle, a 10,000-day future coupled with pledge to resell tomorrow should sell for the same price as a one-day claim on x.

However, x isn't all that is observed as time passes. Time offers opportunities to reappraise history and revise core hypotheses. Call it a change in paradigm or conceptual framework. Replacing EMA Reverters after sustained bad prediction of E^{fut} is a stylized example.

Figure 6.6: RMSEs for E^{fut} using various methods

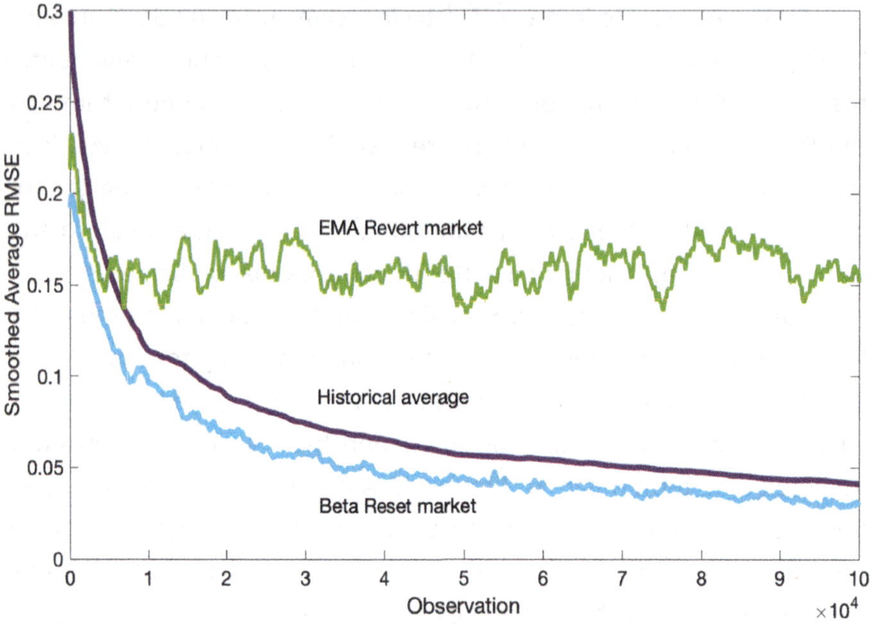

Major reappraisals are sporadic. Their timing depends not just on discoverers' flashes of inspiration but also on others' openness to persuasion. Betting on convergence might justify a long-term hold as that hedges out uncertainty over timing.

A gambler i who expects shift $\Delta\varepsilon_i$ in E^{fut} with variance σ_i^2 and applies Kelly fraction f_i will bet $f_i(\Delta\varepsilon_i - \Delta\varepsilon)/\sigma_i^2$ given consensus shift $\Delta\varepsilon$ with equilibrium

$$\Delta\varepsilon^* = \sum_i \frac{\kappa_i f_i}{\sigma_i^2}\Delta\varepsilon_i \Bigg/ \sum_i \frac{\kappa_i f_i}{\sigma_i^2} . \tag{6.6}$$

This mimics the market for the present claim, with $\Delta\varepsilon_i$ replacing E and σ_i^2 replacing σ_E^2. However, it make a huge difference in substance, since the relevant news concerns beliefs $p(\theta)$ rather than returns x. For example, suppose that Bernoulli E^{fut} is currently priced at 0.45 while we are confident that it is worth 0.5. If our current $E = 0.2$ and we expect that to strengthen the market influence of learners with low long-term μ, we might want to short E^{fut} or at least defer some purchases.

In principle, a well-developed futures market, including options, allows identification of market beliefs in fine detail. Yet that doesn't make it easy to hedge market beliefs and trade solely on fundamentals. The profits flow best to those just ahead of the crowd and watching where it is leaning. These conclusions echo Keynes (1936, Chapter 12), who famously likened securities investment to a beauty contest won by the judges who best predict the consensus choice:

> The energies and skill of the professional investor and speculator are mainly occupied [...] not with making superior long-term forecasts of the probable yield of an investment over its whole life, but with foreseeing changes in the conventional basis of valuation a short time ahead of the general public... This behavior is not the outcome of a wrong-headed propensity. It is an inevitable result of an investment market organized along the lines described.

Momentum Trading

Since markets are volatilely volatile, with small doubts and deviations in higher cumulants entering in ways that are hard to predict, market study encourages far more bets on short-term moves than on their long-term destination. Excess price volatility exacerbates the short-term focus. To see why, let's compare the expected values of short-term and long-term insight for two magical machines:

- Overnight Oracle identifies daily one of 100 iid components q_i of tomorrow's change ΔP in market price.
- Future Prophet identifies a halving or doubling of true long-term value that is expected to correct within T years.

For Overnight Oracle, the optimal Kelly bet per unit wealth generates an expected profit of $\langle q_i^2 \rangle / \text{var}(\Delta P - q_i) \approx 0.01$ per trading day. For Future Prophet, assuming constant log price convergence at yearly rate $\log(2)/T$ and yearly price volatility ς, the optimal Kelly bet per unit wealth generates an expected profit of approximately $(\log(2)/\varsigma T)^2$ per trading year. When $\varsigma = 15\%$, T must be less than three years for Future Prophet to offer more value, whereas if ς were halved the maximum T favoring Future Prophet would double to six years.

Granted, an efficient portfolio strategy could employ both machines. Future Prophet requires little daily maintenance, imposes few transactions costs, and is little correlated with Overnight Oracle. Too bad it can't exist. No one, not even magical machines, can sample the future.

One useful proxy is the capital-weighted market price $F^{fut} \equiv \langle E_i^{fut} \rangle_k$ of the various forecasts E_i^{fut}. Since no reasonable investor can be sure, each E_i^{fut} is associated with a variance V_i^{fut} and some weight on the consensus F^{fut}. While some news affects future x independently of current x, it is hard to predict the impact. The only immediate gauge is the shift in consensus F^{fut}. On observing that, previously agnostic investors usually shift their E_i^{fut} in the same direction, suspecting that others know something they don't. This usually shifts F^{fut} even more but there is one big caveat: too much shift for too long will trigger a counter-reaction that shifts F^{fut} back. Betting that past outperformance persists is known as momentum trading, whereas betting on a rebound is known as mean-reverting or contrarian trading.

Predictable momentum contradicts the FIRE-based efficient markets hypothesis championed by Samuelson (1965) and Fama (1965), where current prices incorporate all available information and render past prices irrelevant. However, the evidence for significant returns momentum is very strong (Jegadeesh and Titman 1993, Carhart 1997, Ilmanen 2011, Moskowitz et al 2012, Asness et al 2013, Lim et al 2018). Fama and French (1996, 2008) described momentum as an "embarrassment" to the efficient markets hypothesis and the "premier anomaly" they could not explain.

Like other violations of FIRE, momentum is often attributed to irrational investor behavior. Barberis et al (1998) stress alternation between underreaction to earnings news and overreaction due to model misspecification. The investor thinks that gains and losses alternate in Markov processes of either short duration or long duration with occasional switches between them, when in fact returns are random with perfectly stable risk. Hirshleifer (2001) offered a more general framework of under/overreaction and folded in overconfidence and herding.

A RATE perspective acknowledges misspecification but explains it more benignly. There are strong incentives to forecast current E well and correct errors quickly. However, E^{fut} is much harder to predict. Errors are rife and hard to correct other than by comparing E_i^{fut} to consensus F^{fut}. This is bound to promote herding that is self-fulfilling in early stages and prone to overshooting in later stages.

From a RATE perspective, momentum should sway growth stocks far more than value stocks due to the higher weight of E^{fut} relative to E. The empirical evidence strongly confirms this prediction (Fama and French 1996, Jegadeesh and Titman 2001, Asness et al 2013). Behavioral finance acknowledges the importance of high E^{fut} weight too but frames it as growth stocks attracting more irrational investors (Daniel et al 1998).

Incentives for financial news and analysis also favor a short-term focus. Each morsel of current news has an immediacy that gives a potential trading edge to those who receive and process it fastest. Distinguishing correct interpretations from misinterpretations takes longer. So does infusing new perspectives from research or refining them through reasoned debate. While this can add more long-term value than a flash news report, it is harder for publishers to monetize. The most notable exceptions involve publications by the research departments of brokers or fund managers, which aim to advertise their superior know-how to clients or to induce others to buy what they already hold.

Immediacy explains why professional sports have suffered far less than professional music from the pirating of recordings. Once the final score is known, full game replays lose most of their excitement and are savored only by coaches and scouts. In contrast, most listeners savor new music better on rehearing.

For equities, every trading moment provides the latest score. Imagine a security that pays dividends tied to the difference between observed global temperatures and historic benchmarks. Its core value would hinge on future global warming. Yet most of the related financial news would discuss today's price action and unusual current weather.

Hence, a host of factors reinforce the market's natural tendency to focus on short-term prediction. Each trader strives for an edge through quick response and better understanding of others' short-run beliefs, with little concern for long-term perspectives. This allows for considerable heterogeneity and offers occasional opportunities for farsighted investors to profit immensely. But it warns against relying on consensus long-term forecasts. Capital markets shine at quick correction of short-term errors, not long-term prophecy.

From a RATE perspective, this is less of a market failure than an unavoidable cost of our dimness about the future. Summarizing extensive empirical research, Tetlock and Gardner (2015) find that the top 2% of human forecasters share the traits associated with Perfect Learners. They entertain multiple hypotheses, evaluate them probabilistically as new evidence is revealed, and seek to correct errors quickly. However, they are best at predicting events less than a year ahead and flail on predictions with long time horizons.

Gabaix (2020) similarly highlights "agents' partial myopia toward distant atypical events" although he frames it as behavioral aberration. In his view, acknowledgment of this myopia is vital to "empirically relevant enrichment of the New Keynesian model". While the analysis falls outside the scope of this book, it illustrates the potential application of RATE to areas well outside finance proper.

7

Risk Premia

We have now addressed four major puzzles of equity markets: excess volatility, highly volatile volatility, hyperactive trading, and momentum. These violations of FIRE are commonly attributed to irrational exuberance that competitive markets ought to dampen. RATE coupled with Kelly gambling provides a far more plausible explanation.

This chapter addresses a fifth major puzzle: the high risk premia for equities. While this was highlighted decades ago by Mehra and Prescott (1985), none of the proffered explanations have fully captured economists' imagination. Ironically, the leading approach can be described as rational cowardice, where traders correctly assess risks but are extraordinarily averse to taking them. How to reconcile this with alleged irrational exuberance is a major challenge for both FIRE and behavioral finance.

The coupling of rational learning and Kelly gambling provides a far more plausible explanation. Most professional traders aim for such coupling. However, the dominant approach among finance theorists rules that out. It treats aggregate equities as claims on aggregate consumption and prices them to reflect consumer utility maximization. Below I review that approach, expose its core limitations, and present a constructive solution focused on equity dividends and earnings.

The best-known models of long-term consumer utility maximization presume time-additive expected utility (EU) with constant relative risk aversion (CRRA). Time-additivity requires

$$EU \equiv \int_0^{\infty} \langle u(c(t)) \rangle e^{-\rho t} dt \tag{7.1}$$

given consumption rate $c(t)$ at time t, utility flow $u(c)$, and time discount rate ρ. CRRA requires

$$u(c) \equiv \frac{c^{1-\gamma}-1}{1-\gamma} \tag{7.2}$$

for $\gamma > 0$ with limit $\log c$ for $\gamma = 1$. CRRA makes relative asset allocations and risk premia independent of total wealth W, a simplification that is nearly indispensable in macro modeling.

To cultivate intuition for the tradeoffs γ implies, imagine a lottery ticket offers a 10% chance of infinite riches. If you would bet everything on that ticket, your implied γ is less than 1. Maximum bets plummet as γ rises. Are you willing to pay half your wealth? Then your implied γ is 1.15. A quarter of wealth? Then your implied γ is 1.37. As Figure 7.1 shows, willingness to pay shrinks from 10% of W at $\gamma = 2$ to 1% of W at $\gamma = 10$.

Figure 7.1: Willingness to pay for 10% chance of infinite wealth for 2≤CRRA≤10

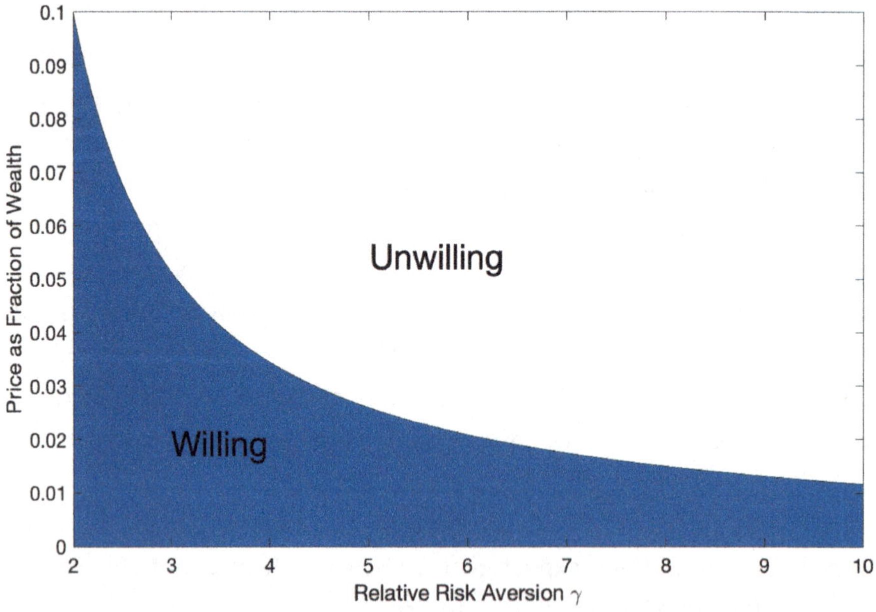

My guess is that most people would pay at least 10% of W, which implies $\gamma < 2$. Much empirical research backs this supposition. Fitting a variety of models, Hansen and Singleton (1983) reported that most γ point estimates fell between 1 and 2. So did Szpiro (1986), who drew on property and liability insurance data for 15 countries. (Szpiro and Outreville (1988) later qualified this conclusion. They reported $\gamma < 5$ for 29 of 31 countries with pooled average 2.5. Outreville (2015) found that risk aversion was negatively correlated with higher education. Better educated countries were associated with $\gamma < 3$ and an average close to 2.)

Using life insurance data for 2400 households, Halek and Eisenhauer (2001) estimated $\gamma \leq 1.5$ for about 70% and $\gamma \leq 2.3$ for about 80%. Using estimated labor supply elasticities from 33 studies, Chetty (2006) found an average $\gamma \approx 1$ and showed that $\gamma > 2$ "would require that wage increases cause sharper reductions in labor supply than estimated in any of the studies". Using six very different surveys on the marginal utility of income, Layard et al (2008) estimated $1.19 \leq \gamma \leq 1.34$. Using data on wealth and portfolio allocations of 8000 Italian households, Chiappori and Paiella (2011) estimated a median γ of 1.7. Using Gallup Poll data on self-reports of personal well-being in 80 countries, Gandelman and Hernandez-Murillo (2015) could not reject the null hypothesis $\gamma = 1$.

In contrast, nearly all papers that aim to account for equity risk premia require much higher γ. Their identification of equity investors with representative consumers, which in turn identifies aggregate equity dividends with aggregate consumption, leaves them no choice. Unless agents are highly risk averse, observed variations in consumption don't deserve nearly as much insurance as equity risk premia provide. As first noted by Rietz (1988), anticipation of rare disasters can raise the risk premium associated with a given γ, and subsequent work by Barro (2006) and others indicates that the effect is substantial. However, their empirical fits require a γ of at least 4.

For more insight, let us examine the crisp solution Martin (2013) derived using the cumulant generating function K of the growth Δx of log consumption. Note that $\mathrm{K}(1) = \log\langle e^{\Delta x}\rangle$, the mean growth rate expressed in log terms. Given (7.1) and (7.2), the dividend ratio Q^{-1} is

$$Q^{-1} = \rho - \mathrm{K}(1-\gamma), \tag{7.3}$$

while the risk premium r_p for equities over risk-free bonds is

$$r_p = \mathrm{K}(1) + \mathrm{K}(-\gamma) - \mathrm{K}(1-\gamma). \tag{7.4}$$

When consumption growth is log-normal with volatility σ, $r_p = \gamma\sigma^2$, which would require $\gamma > 10$ to fit the evidence. To justify the same r_p at much lower γ, $\mathrm{K}''(\tfrac{1}{2}-\gamma)$ must be highly positive, say due to a 1%+ chance of a 30%+ plunge in consumption. Historical disaster risks as reported by Barro and Ursúa (2008, 2012) are broadly consistent with that and suggest $\gamma \approx 4$. However, roughly half of big drops are subsequently reversed, and after adjusting for this Nakamura et al (2013) estimated $\gamma \approx 6$.

Note that Q in (7.3) is constant. It never rises on anticipation of high growth or falls on fears of contraction. Price volatility always matches dividend volatility. That's because Martin, following a common practice in such models, compresses a regime-switching process into repeated single-period iid outcomes that no investors can adjust to mid-course. Beliefs correctly anticipate the long-run distribution but never gain more specific information and never change. As we will see, relaxing these restrictions makes high CRRA even harder to defend.

Let us return to the Markov-switching model of equities examined in Chapter 4. To recall, $\bar{Q}(\mu)$ denotes the fair price-to-dividend ratio for a known log dividend drift μ, while Q_π denotes the mean $\bar{Q}$ just after reset at rate λ to a stationary distribution π. With risk neutrality, these must satisfy the one-instant look ahead condition (4.11), which can be rewritten

$$\bar{Q}(\mu) \approx 1 \cdot dt + e^{-\rho dt} \langle e^{dx} \rangle \left((1-\lambda dt)\bar{Q}(\mu) + \lambda dt \cdot Q_\pi \right). \tag{7.5}$$

for discount rate ρ. For time-additive CRRA utility, the only change to valuation is that each multiplier e^{dx} on growth gets weighted by its marginal utility $e^{-\gamma dx}$. This replaces the expectation $\langle e^{dx} \rangle$ in (7.5) with

$$\left\langle e^{(1-\gamma)dx} \right\rangle \approx 1 + (1-\gamma)\mu dt + \tfrac{1}{2}(1-\gamma)^2 \sigma^2 dt = 1 + \mathrm{K}_B(1-\gamma|\mu,\sigma),$$

where $\mathrm{K}_B(s|\mu,\sigma) = \mu s + \tfrac{1}{2}\sigma^2 s^2$ denotes the cumulant generating function for Brownian motion. The solution modifies (4.11) to

$$Q = \frac{\langle q(\mu) \rangle}{1 - \lambda \langle q \rangle_\pi} \quad \text{for} \quad q(\mu) \equiv \frac{1}{\lambda + \rho - \mathrm{K}_B(1-\gamma|\mu,\sigma)}. \tag{7.6}$$

All feasible solutions require $q > 0$ and $\langle q \rangle_\pi < 1/\lambda$. The limiting cases as $\lambda = 0$ or $\lambda \to \infty$ match (7.3) although neither induces a non-normal K.

While (7.6) is admirably crisp, it cannot be reconciled with real-life equity market behavior. We need disaster risks with sufficiently negative μ, low λ, and high γ to justify a substantial risk premium. Yet if $(\gamma - 1)\mu$ gets too negative, $q(\mu)$ will be negative. For the best-fit two-drift model discussed in Chapter 4, no feasible solution with $\rho < 0.04$ allows $\gamma > 3$. No gamma-reset model of crisis allows $\gamma > 1$ since crisis μ is unbounded.

Furthermore, any $\gamma > 1$ in (7.6) makes Q fall as growth prospects improve. As a result, equities will be less volatile than dividends in normal times and soar in crisis when dividends plunge. The contrasts are glaring. With $\rho = 0.01$ and $\gamma = 2$, $\bar{Q}(\mu_L)$ for the best-fit two-drift model is about twice $\bar{Q}(\mu_H)$. Over 80% of Q lie within one point of their minimum with price volatility ς of less than 5%. Q is highest when ς exceeds 40%. Figure 7.2 charts a 100-year simulation. No common equities behave remotely like the safe-haven asset it depicts.

We can gain wiggle room by replacing (7.1) with recursive Epstein-Zin (1989) preferences that separate risk aversion at a given time from consumption/savings preferences over time. The intertemporal elasticity of substitution ψ is high when people eagerly defer gratification to earn more in the future and low when people don't. Under time-additive CRRA utility, $\psi = 1/\gamma$, so high risk aversion drives attachment to current consumption. To make equities more pro-growth, we need $\psi > 1$.[19]

Empirically, there is much less consensus on plausible ψ than there is on plausible γ. However, most estimates place ψ well below 1. Reviewing 2735 estimates reported in 169 published studies, Havranek (2015) found that researchers discarded negative and insignificant estimates too often. The corrected mean estimate of ψ was 0.3-0.4. While Havranek et al (2015) found higher ψ estimates in rich countries with developed equity markets, relatively few exceeded 1.

Figure 7.2: Q and ς simulation when equities pay aggregate consumption to investors with CRRA=2

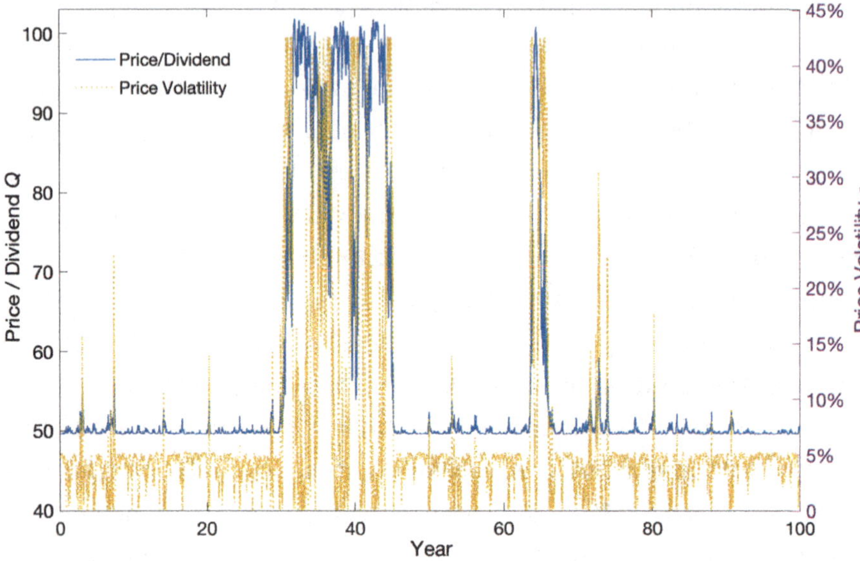

Market behavior looks more realistic when dividends are modeled as ℓ-times leveraged consumption proxied by a power function ac^{ℓ}. This replaces $\gamma-1$ in (7.6) with $\gamma-\ell$. Q flips back into positive correlation with dividends for $\ell>\gamma$ and we regain the risk-neutral solution when $\ell=\gamma+1$. For more flexibility, time-additive utility can be replaced with recursive utility and investments can be allowed in risk-free bonds. As Wachter (2013) and Tsai and Wachter (2017) have shown, this can generate far more verisimilitude than (7.6) can. Still, the results seem fragile. Good fits typically leave both γ and ψ implausibly high. The power function approximation to leverage fails badly for high ℓ. The equilibria are not robust to modest changes in the parameters. They also downplay the market's lags and errors in identifying ever-changing trends.

In short, no one has managed to reconcile high equity risk premia with empirical data on aggregate consumption, γ and ψ.[20] That includes an imaginative effort by Weitzman (2007), who highlighted the uncertainty in estimating the mean growth of consumption. Given a presumed Gaussian distribution, the classic estimator has a heavy-tailed t-distribution, where only cumulants of order less than the sample size are finite. If that is what observers apply, fair Q and r_p are infinite and any market price is too low rather than too high. While this nominally resolves the risk premium puzzle, it poses bigger puzzles about why there is any trading in equities or clearly defined prices. Bakshi and Skoulakis (2010) traced the problem to the unbounded potential losses implied by a t-distribution. When these are excluded, Weitzman's core claim collapses.

Is there any viable way to tie equity prices to aggregate consumption risk? Here are two reasons for pessimism:

- Equity investors are highly unrepresentative consumers. They are far richer than average, have ample cushion against downturns and expect to leave large endowments to others.
- Equities are claims on net corporate profits, which do not embrace most of GDP or even a stable share. According to the US Bureau of Economic Analysis (2024), US corporate after-tax profits since World War II have averaged less than 8% of GDP, with large cyclical variations and evidence of changing secular trends. See Figure 7.3.

Figure 7.3: Corporate after-tax profits as share of quarterly US GDP 1947-2023

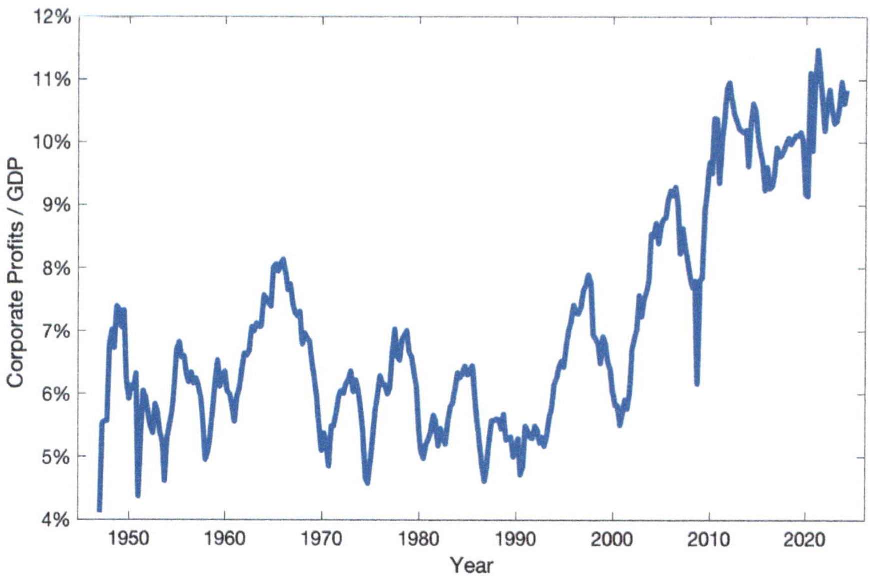

Gambling Perspective

To better account for risk premia, we need to drop all pretenses that equity dividends mirror aggregate consumption and that equity investors mirror average consumers. Real-life equity markets are dominated by very rich people and institutions trying to get richer. Much of their consumption is insulated from equity gains or losses and responds only marginally to wealth. What they most value is the influence their wealth emanates. To this end they hire investment experts and entrust them to manage risk-reward tradeoffs wisely. While professional investment managers are not free of behaviorist foibles or conflicting incentives, I will ignore those effects here.

This brings us back to the treatment of equity markets as casinos of Kelly gamblers, which Chapter 5 introduced. When gamblers bet only against each other, net holdings must be zero, in which case the risk premium must be zero too. Net holdings will be positive if and only if the risk premium is positive.

To ascertain the relation between risk premia and net longs, let us start with a market of Perfect Learners. They all think alike, with no trading between them, so we can view the market as a single holder. Given a bet on w equities priced at P each with known mean rate of profit Π and volatility ς, the expected log rate of return is

$$w\Pi P - \tfrac{1}{2} w^2 \varsigma^2 P^2 , \tag{7.7}$$

This is maximized when the value $wP = \Pi/\varsigma^2$, the full Kelly criterion. As noted in Chapter 5, allowing for uncertainty in Π or aversion to drawdowns argues for betting closer to half Kelly.

For an alternative justification of fractional Kelly, suppose the Perfect Learner behaves like a CRRA investor. Expected CRRA utility will be maximized to first order at investment $\Pi/(\gamma\varsigma^2)$. This suggests setting $f \approx 1/\gamma$ (see Kadane (2011) for further discussion), which empirically lies between 0.5 and 1.

Let us now allow heterogeneity in views while maintaining a common f. Let P_i denote the gambler i's view on the future expected value and let $P_E \equiv \sum_i k_i P_i$ denote the market consensus view. Given a market price P_F, each fractional Kelly bet over the next instant dt is $fK_i(P_i - P_F)/(P_F^2 \varsigma^2 dt)$ for an aggregate bet of $fK(P_E - P_F)/(P_F^2 \varsigma^2 dt)$. In equilibrium this must match the N equities outstanding, so

$$\frac{P_E - P_F}{P_F} = \frac{NP_F}{fK} \varsigma^2 dt . \tag{7.8}$$

The risk premium rate $r_p \equiv \varsigma^2 NP_F/(fK)$ is the product of three factors: the reciprocal of aggregate f, the share of aggregate equity value NP_F in total speculative capital K, and price variance ς^2. Since ς^2 is far more volatile than the other two factors, (7.8) implicitly predicts a highly negative correlation between ς and P for major equity indices. The correlation should be especially strong in highly turbulent markets due to the dependence on volatility squared.

This is readily confirmed. The correlation between the S&P 500 index SPX and the VIX index of 30-day implied volatility on the SPX is extremely negative, to the point of calling VIX a "fear gauge". Some of this reflects the higher average uncertainty during bear markets that was noted in Chapter 4. However, the coupling of big spikes in VIX with big drops in SPX seems best explained by (7.8). Carr and Wu (2006) found that correlations of -0.6 in quiet times surged to nearly -0.9 in crises.

If we identify K with total capital NP_F invested in equities, then

$$r_p \approx \varsigma^2 / f . \tag{7.9}$$

This ties r_p closely to price volatility ς rather than the lower volatility σ of dividends or consumption. Since ς often exceeds 15%, (7.9) can easily account for risk premia of several hundred bps even when risk aversion is low. Furthermore, since ς tends to rise when dividends fall, fluctuations in r_p make Q fluctuate even more, which boosts average ς. French et al (1987) confirmed a positive relation between r_p and ς.

Judging from (7.9) and our estimates of f, the mean excess return $\langle R_{eq} \rangle$ for equities over short-term Treasury bills should be on the order of 1-2 times $\langle \varsigma^2 \rangle$. Figure 7.4 offers some empirical support. Drawing on data from Barro and Ursúa (2012, Table 2), it compares $\langle R_{eq} \rangle$ and $\langle \varsigma^2 \rangle$ for both OECD and non-OECD countries, covering both 1879-2009 and the sub-period 1946-2009. For the OECD, the two measures nearly match in both periods, which at first glance suggests $f \approx 1$. However, the variances are computed using yearly data and hence incorporate both $\langle \varsigma^2 \rangle$ and a variance of yearly means. Correction would lower the estimated f.

For non-OECD countries, $\langle R_{eq} \rangle$ is roughly half of variance in both periods, so at first glance their risk premia appear unduly low. However, most non-OECD equity markets are far less developed than their OECD counterparts and far more influenced by foreign investors. Much of their risks can be diluted through diversification. Consequently, their excess returns are better compared to OECD variances than to their home market variances.

Figure 7.4: Average excess yearly return versus variance for equities in OECD and non-OECD countries 1870-2009

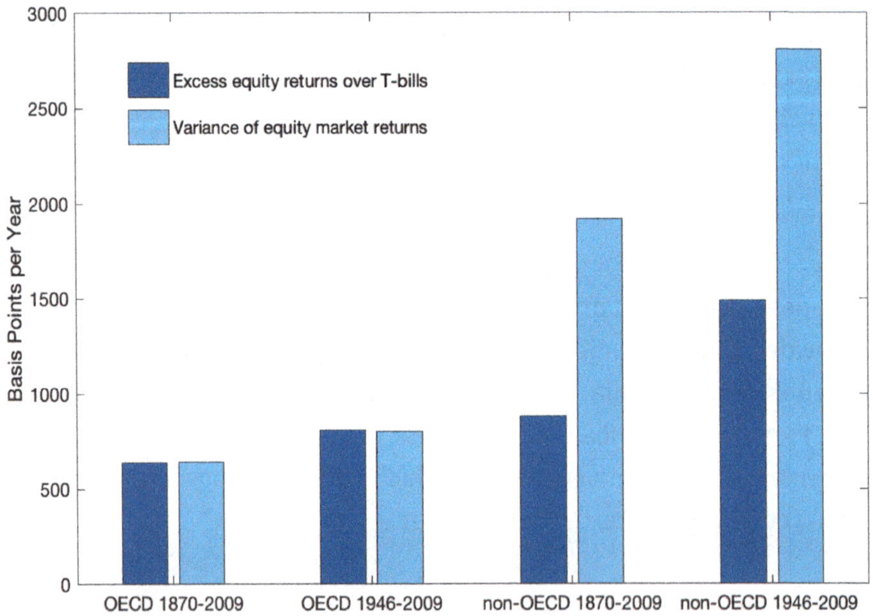

For the United States, Mehra and Prescott (2003) estimated an average r_p of 692 bps for 1889-2000. Bansal and Yaron (2004) estimated 633 bps for 1929-1998. Siegel (2022, Chapter 2) estimated an average r_p of 430 bps relative to short-term bonds for 1802-2001. All these estimates are based on ex-post returns, so the unexpectedly good performance of the US equity market would overstate them. Fama and French (2002) contended that the ex-post 557 bps average excess return on the S&P for 1871-2000 overstates r_p by 125-200 bps. They estimated r_p from projections of dividend and earnings trends and attributed the gap to unanticipated declines in discount rates.

For comparison, the annualized variance $\langle\varsigma^2\rangle$ of daily S&P 500 returns for 1928-2024 averaged 358 bps. All the r_p estimates cited above fall as conjectured within 1-2 times the estimated $\langle\varsigma^2\rangle$. Since $r_p > \frac{1}{2}\langle\varsigma^2\rangle$ is needed to keep mean log returns higher on stocks than on purportedly safe bonds, the excess log return averages 180-500 bps.

Claims on Dividend/Earning Mix

Broad conformance with (7.9) does not guarantee rationality. If equities are systematically too volatile relative to dividends, as was first claimed by Shiller (1989, 2000) and is still widely believed, then all the comparison achieves is the identification of a common excess. Shiller (2025) maintains an online database that allows us to investigate the claim directly. The series "real dividends" consists of aggregate S&P dividends deflated by the Consumer Price Index. Monthly estimates are available from 1871, although they are interpolated from quarterly data. Figure 7.5 charts them on a log scale. It also charts the corresponding series of real earnings.

Dividends became noticeably smoother after 1950. Volatility (annualized standard deviation of monthly log changes) averaged 6.4% for 1871-1950 and 2.3% for 1951-2024. Earnings have been bumpier but show milder secular trends; their volatility has averaged 9.5% apart from the troughs around 1920 and 2009. Dividends averaged roughly 65% of earnings through 1950 but since then the relative gap has significantly widened.

Figure 7.5: Real monthly dividends and earnings of S&P Composite per Shiller

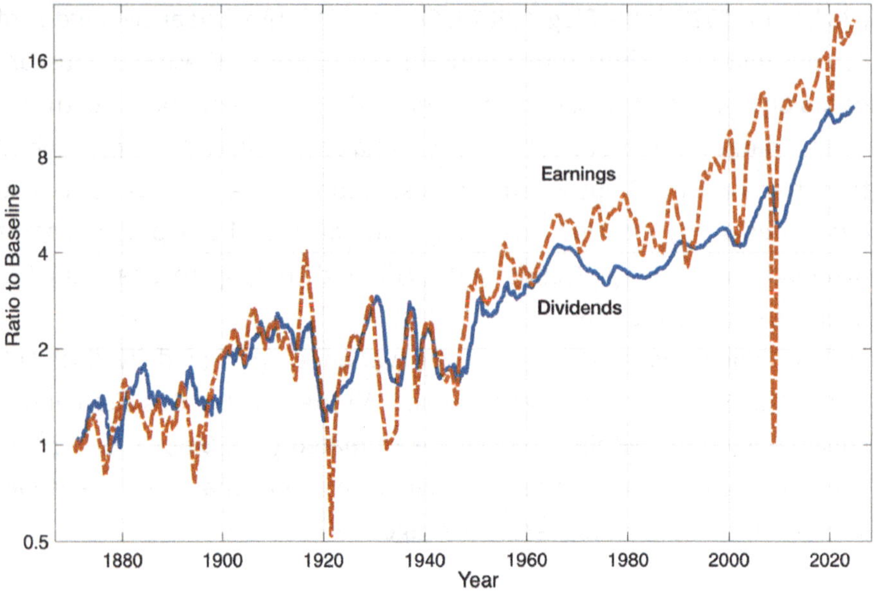

Since firms channel some earnings toward taxes and investment, and since earnings surprises help signal the expected profits from investment, dividends tend to be lower and less volatile than earnings. Managerial discretion promotes additional smoothing. Dividend stability helps reassure lenders and investors of financial health (Lintner 1956, Brav et al 2005). Dividends are often tuned toward projected longer-term earnings, which are highly correlated with stock prices (Marsh and Merton 1987). Larger, older firms tend to smooth more aggressively (Leary and Michaely 2011). These behaviors have evidently strengthened over time.

However, the main changes in aggregate dividend behavior appear to reflect two other trends. One is the increasing importance of fast-growing tech firms that reinvest nearly all their post-tax earnings (Fama and French 2001). The other is the replacement of dividends with share repurchases (Jagannathan et al 2000). The correlation between real dividends and real earnings, excluding the two big troughs, shrank from 0.50 for 1871-1950 to 0.17 for 1951-present and turns negative when the 2009 trough is included.

These shifts significantly dilute the connection between actual dividends and the ideal accruals e^x that markets track (Cochrane 1994, Chen et al 2012). As a first pass at compensation, I proxy x by log real dividends through 1950 and a 75%/25% mix of log real dividends and log real earnings since 1951, with earnings rebased to avoid a spurious jump. Its annual log return averages 0.0167 with aggregate volatility 7.4%.

To estimate two-state Markov switching parameters, I apply the Python module *hmmlearn* to the non-interpolated quarterly data. Like most such packages, it assumes that switches occur only at times of measurement. This understates both the continuous reset rate λ and the $\mu_H - \mu_L$ gap. To convert quarterly estimates to instantaneous estimates I proceed as follows:

- Given quarterly transition rates ${}_q\lambda_{HL}$ and ${}_q\lambda_{LH}$ summing to ${}_q\lambda$, the long-term probability π_H of high-drift is ${}_q\lambda_{HL} / {}_q\lambda$.
- The smallest eigenvalue of the quarterly transition matrix is $1 - {}_q\lambda$. An instantaneous transition matrix can generate that eigenvalue only if $\lambda = -4 \cdot \log(1 - {}_q\lambda)$.
- A good approximation for μ_H and μ_L presumes at most one switch in a quarter, where the new drift lasts on average half a quarter. This widens the estimated difference in drifts by roughly $50\,{}_q\lambda$ percent. The exact answer is given by

$$
\begin{aligned}
\phi &= (\lambda / {}_q\lambda) - 4 \\
\mu_H &= 4 \cdot {}_q\mu_H + \phi \pi_L ({}_q\mu_H - {}_q\mu_L) \, . \\
\mu_L &= 4 \cdot {}_q\mu_L - \phi \pi_H ({}_q\mu_H - {}_q\mu_L)
\end{aligned}
$$

- Annualized $\sigma \approx 2\,{}_q\sigma$ as *hmmlearn* estimates conditional volatility without direct reference to the variance of drifts.

For the dividends/earnings mix, the best fit for $(\pi_H, \lambda, \mu_H, \mu_L, \sigma)$ is $(0.918, 1.100, 0.0463, -0.317, 0.0603)$. The High regime lasts 11.1 years on average with expected log growth of 0.51. The Low regime lasts 1.0 year on average with expected log contraction of 0.31. Extending the 75%/25% mix back to 1871 yields similar results. This coupling of long gradual growth with occasional sharp retreats echoes the Barro (2006) findings for consumption and GDP although the changes in dividend regimes are more frequent and more severe.

On the pricing side, replacing a constant r with the risk-adjusted $\rho+\varsigma_p^2/f$ changes the equilibrium price relations (7.5) to

$$Q_p \approx 1\cdot dt+\left(1-\left(\rho+\varsigma_p^2/f\right)dt\right)\left(1+\left(E_p+\tfrac{1}{2}\sigma^2\right)dt\right)\left\langle Q_{p+dp}\right\rangle,$$

for $p\equiv p(\mu_H)$ and $E_p=\mu_L+p(\mu_H-\mu_L)$. This implies

$$(\rho+\varsigma_p^2/f-E_p-\tfrac{1}{2}\sigma^2)Q_p\approx 1+\left\langle dQ_p\right\rangle/dt. \tag{7.10}$$

Individual beliefs continue to be updated using the Core Learning equation. For two states this is given by (4.14), which is rewritten below as

$$dp=p(1-p)\frac{\mu_H-\mu_L}{\sigma^2}\left(dx-E_p dt\right)+\lambda\left(\pi_H-p\right)dt. \tag{7.11}$$

Let $\hat{\mathbf{Q}}$ denote the vector of candidate $\hat{Q}_p$. If we discretize p over a fine grid and express each $\langle d\hat{Q}_p\rangle/dt$ as a vector product $\Pi_p\hat{\mathbf{Q}}$ using (7.11), we can iterate toward a solution by calculating the various

$$\hat{\varsigma}_p=\sigma+\sqrt{\mathrm{var}(\hat{Q}_{p+dp})}/\hat{Q}_p \tag{7.12}$$

where var denotes variance, solving the linear system over p of

$$\left(\rho+\hat{\varsigma}_p^2/f-E_p-\tfrac{1}{2}\sigma^2\right)\hat{Q}_p-\Pi_p\hat{\mathbf{Q}}=1 \tag{7.13}$$

for a new $\hat{\mathbf{Q}}$, and redoing (7.12) and (7.13). A constant $\hat{\varsigma}$ seed can generate a partially negative $\hat{\mathbf{Q}}$; once this is replaced with $|\hat{\mathbf{Q}}|$ the system converges quickly. However, the mean $\langle\mathbf{Q}\rangle$ is quite sensitive to estimates of rare disasters, consistent with claims of Barro (2006). For stability, I used millions of simulated observations and grouped them into 1000 equally sized clusters for estimation of the transition matrix Π.

Figure 7.6 displays a 40-year simulation using the best-fitting parameters and $f=0.75$, which roughly corresponds to a CRRA $\gamma=1.33$. In a 200,000-year simulation, ς ranged from 6% to 68% with mean 14.1% and root mean square 18.6%. The latter nearly matches the average 18.9% for the S&P since 1928. Q ranged from 25 to 38 with mean 36.9.

Figure 7.6: 40-year simulation of Q and ς when equities discount dividends/earnings mix and risk premia are endogenous

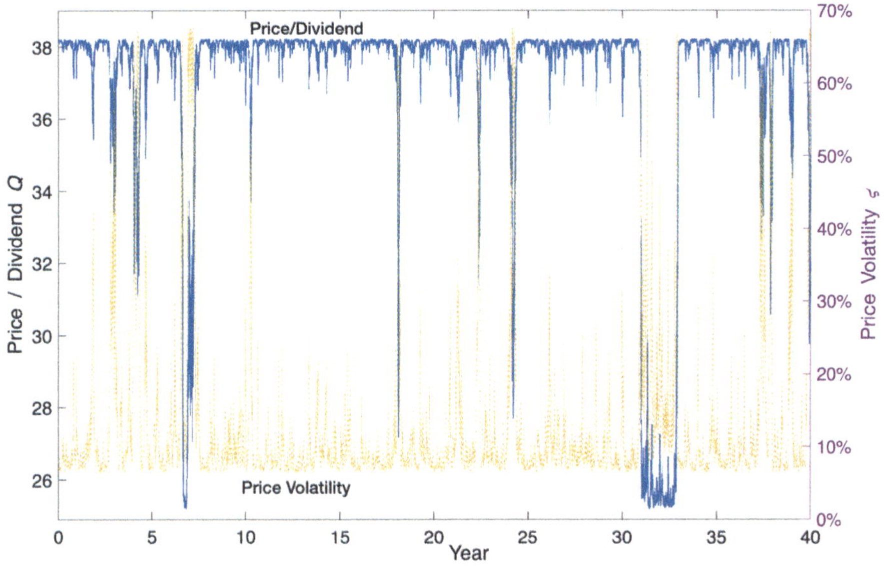

These simulations look far more realistic than any charted previously, not to mention the standard presumption of stable Q and ς. The means are plausible, the variations are plausible, volatility is negatively correlated with business cycles, and GARCH behavior is pronounced without multi-year stretches of consistently low volatility. The precise means and bands can easily be adjusted. Doubling ρ reduces Q by nearly a quarter. Reducing f to 0.5 more than halves Q while raising f to 1 more than doubles Q. None of these changes have much impact on ς.

Figure 7.7 charts the risk premium r_p against Q. It exceeds 10% over most of the range and peaks at over 60%. As the underlying histogram makes clear, Q clusters toward the edges where r_p is much lower. About half are less than 110 bps, 70% less than 200 bps, and 80% are less than 300 bps. However, the average r_p is 465 bps, thanks largely to a top 6% that exceed 2000 bps. The rare peaks are so high that they are bound to appear as bursts of panic, where investors briefly lose their rational minds. The price gyrations then get blamed on panic, when in fact the risk premia and prices respond reasonably to the evidence.

Figure 7.7: Risk premium r_p versus Q overlaid on histogram of Q

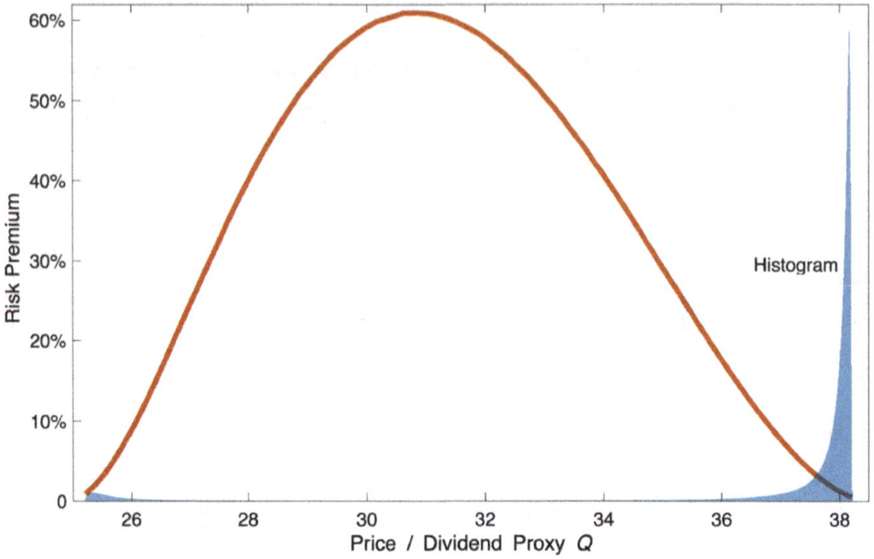

How much does the volatility of volatility contribute to the risk premium? Judging by the gap between the price variance $\langle \varsigma^2 \rangle$ and the squared mean volatility $\langle \varsigma \rangle^2$, the volatility of volatility accounts for nearly half of r_p. It follows that both volatility and the volatility of volatility should be costly to insure but with most of the costs concentrated in crises.

This is well documented empirically. The implied volatility of the VIX index generally exceeds the realized volatility, with especially large gaps in crises (Carr and Wu 2009, Bollerslev et al 2009, Drechsler and Yaron 2011). Bollerslev et al (2011) inferred from the gaps an average investor risk aversion $\gamma \approx 1.82$, which implies $f \approx 0.55$.

To summarize, half-to-full Kelly investor response to earnings-related news seems to account for the main features of equity risk premia, although it is not clear what news investors most respond to with what Kelly fraction f.

8

Uncertain Safety

Let us turn now from equities to debt. From a modeling perspective, debt markets are casino games in which one outcome is far more likely than the rest. That outcome, called servicing, pays what was promised. Any failure, called default, disrupts all future payments. If lenders charge interest of R for a one-year loan that is certain to be repaid and the chance that nothing will be repaid is θ, a pledge to repay \$1 next year is worth $P=(1-E)/(1+R)$ dollars today for $E\equiv\langle\theta\rangle$. In practice, defaulted debt usually catalyzes some repayment later, known as salvage value. If salvage value is worth a fraction $1-\eta$ of the nominal pledge, then

$$\langle\theta\eta\rangle=1-(1+R)P\,. \tag{8.1}$$

This measure is known as the credit spread. When multiple payments are pledged at different times, computation is harder but the same principles apply. Like interest rates, credit spreads are typically reported on an annualized basis.

Merton (1974) pioneered the treatment of debt as a call option on equity. Default is triggered by equity price diffusing below a threshold, which makes default risk a measurable proximity. While empirical evidence strongly rejects the basic model (Jarrow et al 2003), Duffie (2011) traced many violations to "a common unobserved source of default risk that increases default correlation and extreme portfolio risk". By modeling this as a stochastic mean-reverting "frailty factor" and tuning the parameters, Duffie fit the evidence much better.

Since I posit a similar factor, much of what I say about spreads can likely be incorporated into Duffie's framework. My focus is the question of what can or cannot be learned from news on debt servicing. We will see that doubt about θ looms large, partly because much evidence is needed to ascertain small default risks and partly because older evidence loses relevance too quickly. The resulting turbulence has some parallels to equity market turbulence and some distinctive features.

The reputedly safest borrowers, typically with $\theta < 50$ bps, are called investment grade. Markets for investment-grade debt are massive. About \$18 trillion of global corporate debt is rated investment grade (S&P Global 2023). Add to that some \$27 trillion of US Treasuries, \$8 trillion of Japanese government bonds, and \$3 billion each of British, German, French, and Italian bonds, all of which are rated investment grade. Daily trading of investment-grade debt now averages over \$1 trillion.

Since investment-grade defaults are rare, credit spreads rarely look right. Mostly they seem too fearful and occasionally not nearly fearful enough. Reliance on averages is suspect since no debt from different issuers is perfectly equal in risk and no single issuer's default risks are perfectly constant over time. Yet no traders act as if the risks cannot be quantified. The very practice of quoting credit spreads in basis points testifies to close attention. It is a mammoth rebuke to Knightian/Keynesian assertions about financial uncertainty.

Investment-grade credit spreads also rebuke FIRE, which treats beliefs as near-singletons. For example, the most trusted bonds, rated triple-A or a slightly lower double-A, are deemed to have annual default risk of less than 10 bps. Yet there is no way to precisely identify the risk.

For perspective, consider ratings of pharmaceutical safety. Suppose a new medicine has been tested on T patients with no instance of harm. How safe is it? If the true risk were $4/T$, T iid trials would be 50 times likelier to cause some harm than to cause none. Nevertheless, medical regulators are so cautious that they typically refuse to certify the risk as less than $4/T$. If we apply the same standard to bonds, we need 4000 iid yearly observations without default to justify high confidence that yearly θ is less than 10 bps. How can we find that many similar comparators for a reputedly safe debtor like the US or Japan?

In practice, the low risks are largely inferred from defaults on higher-risk credits. Formal credit ratings implicitly treat credits as if infused with various amounts of a common "riskium" element. The more riskium, the greater the default risk. Yet the appropriate scaling is uncertain, particularly for tiny risks. The only way to justify high confidence is to tap a long history of iid-quality comparators. If we find effectively D_{eff} relevant defaults in T_{eff} years, we can estimate $E = D_{eff}/T_{eff}$ with a relative standard deviation of about $1/\sqrt{D_{eff}}$. For 95% confidence that θ lies within 50% of E, we need 16 relevant defaults. That requires roughly 3,200 years of observations for $E = 50$ bps and 16,000 years for $E = 10$ bps.

While humans have great abilities to infer relevance, the empirical evidence tells against high confidence. If benchmark θ were stable, E should have gradually converged to it, with current estimates changing at most a few bps per year. That is nowhere close to what we observe.

Figure 8.1 charts average daily credit spreads since 1997 for double-A US corporate bonds as computed by ICE BofA Data Indices (2024a). The only plausible explanation for the fluctuations is that spreads adjust to evidence that baseline θ is changing. The quarterly changes are often quite large, which suggests a high weight on recent news. Yet with only a few thousand investment-grade corporates, there cannot be enough evidence to track θ with high certainty.

For simplicity, I will assume that servicing is continuous until default and that default is followed immediately by issuance of similar new debt. I will call this irregular cycle a credit. The default risk θ over a short interval dt is approximately νdt, where ν denotes the instantaneous default risk. Given subjective probability density p with $E \equiv \langle \nu \rangle_p$, Bayes' rule requires

$$\begin{array}{ll} p(\nu) \to \dfrac{p(\nu)\nu}{E} & \text{on default} \\ \dfrac{dp(\nu)}{dt} = p(\nu)(E - \nu) & \text{otherwise} \end{array} . \tag{8.2}$$

Figure 8.1: Credit spreads for double-A US corporate bonds

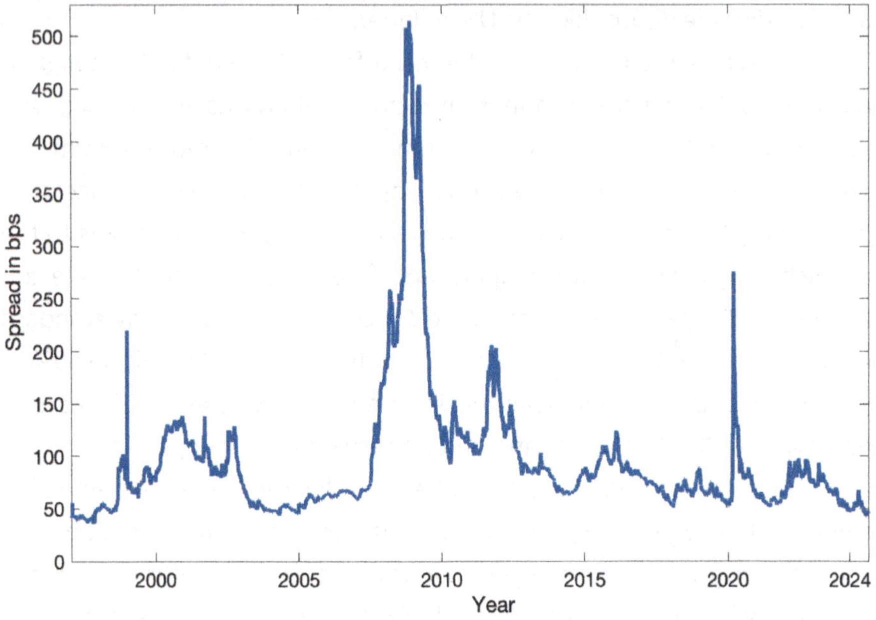

Recall our definition (3.2) of the learning rate $\eta \equiv 1/T_{eff} \equiv V/E$. The counting method mentioned earlier records D_{eff} effective defaults in T_{eff} effective years and estimates $E = D_{eff}/T_{eff}$. We can reconcile this with (8.2) by defining $D_{eff} \equiv E^2/V$, in which case

$$\begin{aligned} dE &= V/E = \eta = 1/T_{eff} = E/D_{eff} && \text{on default} \\ \frac{dE}{dt} &= -V = -\eta E = -E/T_{eff} && \text{otherwise} \end{aligned} \tag{8.3}$$

The asymmetric response to default and servicing is a hallmark of debt markets, although it is very noticeable only when D_{eff} is small.

In the standard counting routine, D_{eff} jumps by one at every default while T_{eff} increments at the rate of one per credit per year. This is rational provided ν never changes and p is gamma-distributed with shape D_{eff} and rate T_{eff}. At unit shape, a gamma density is exponential, with decay rate $1/T_{eff}$. As defaults accumulate, gamma densities look increasingly lognormal. For $D_{eff} < 1$, gamma densities are infinitely steep at the origin but have a long, relatively flat tail. See Figure 4.3 for illustrations.

While no uncertainties about default are exactly gamma-distributed, the dominant shape tends to be gamma-like.[21] Absent doubts about stability, Bayesian updates will quickly drive D_{eff} well above 1 and rule out big jumps in E on news of default. Yet we know that tiny E occasionally soar in crisis, when contagion risks are perceived high. We also know that D_{eff} must be low for highly trusted debt when relevant sample sizes are small. For example, if $T_{eff} < 600$ years for Japanese government bonds, any $E < 5$ bps implies $D_{eff} < 0.3$.

D_{eff} cannot be less than 1 without concentrating most density well below the mean. It behaves somewhat like a two-point distribution that places $1/(1+D_{eff})$ weight on $\nu = 0$ and the rest on $\nu = E + E/D_{eff}$. Since default crushes beliefs closest to zero, it potentially multiplies E manyfold without necessarily implying complacency before or overreaction after.

For D_{eff} to occasionally drop substantially, much less to fall below 1, (8.2) must be supplemented with doubts about stability of ν. Let us also allow for a pool of N iid credits servicing at the same time. The revised learning equation per instant dt is

$$\begin{array}{l} p(\nu) \to \dfrac{p(\nu)\nu}{E} \quad \text{on default} \\ dp(\nu) = Np(\nu)(E-\nu)dt + \langle \text{net migration rate into } \nu \rangle dt \end{array} . \tag{8.4}$$

We can relax this formulation to mix credits of different risk, provided we treat each credit as a bundle of riskium that defaults if even a single riskium component fails. While only the aggregate rates $\sum_i \nu_i$ and $\sum_i E_i$ matter for updates, we will focus on the average value $\nu \equiv \sum_i \nu_i / N$ and $E \equiv \langle \nu \rangle$. This transforms (8.3) into

$$dE = \begin{cases} V/E = 1/T_{eff} = E/D_{eff} & \text{on default} \\ (-NV + E^{shift})dt & \text{otherwise} \end{cases} . \tag{8.5}$$

To extend this, let us return to (4.4) and replace dy/σ with $1/E$ on default and $-Ndt$ otherwise. On default, the moment generating function

M shifts to M'/E. Take the logarithm to obtain the new cumulant generating function K. Otherwise $d\mathrm{K} = d\mathrm{M}/\mathrm{M} = N\left(E - \mathrm{M}'/\mathrm{M}\right)dt$ plus an adjustment $\mathrm{K}^{shift}dt$ for expected switches. To summarize,

$$d\mathrm{K}(s) = \begin{cases} \log(\mathrm{K}'(s)) - \log E & \text{on default} \\ \left(NE - N\mathrm{K}'(s) + \mathrm{K}^{shift}(s)\right)dt & \text{otherwise} \end{cases}. \tag{8.6}$$

The predicted skewness for small E^{shift} is approximately $1/\sqrt{NE}$. In practice this will be tempered by intimations of default that circulate before the actual event. However, some positive skewness ($\kappa_3 > 0$) or related asymmetry should be evident. Here is some confirmation:

- For 14 series of average US corporate credit spreads from 1997 through 2024, subdivided by credit grade or duration (ICE BofA Data Indices 2024a), daily declines consistently outnumbered daily rises. Equal likelihood is rejected with confidences of 99.99% for 10 series, 99.98% for 3 series, and 99.4% for triple-A spreads.
- For the same series divided into six-month intervals, intervals with $\kappa_3 > 0$ consistently outnumbered intervals with $\kappa_3 < 0$. Equal likelihood is rejected with confidences of 99.99% for 3 series, 99.98% for 4 series, and at least 99.4% for 6 other series. The outlier was 97.6% for bonds with maturities of 7-to-10 years.
- Similar results are obtained for 7 series of non-US corporate credit spreads from 1998 through 2024 (ICE BofA Data Indices 2024b). For every series, equal likelihood of daily rise versus decline is rejected with over 99.99% confidence. while equal likelihood of positive or negative κ_3 is rejected with over 99% confidence.
- For older data, Longtermtrends (2024) has estimated monthly US triple-A spreads by subtracting 10-year Treasury yields from yields on trusted long-maturity corporate bonds. Equal likelihood of monthly rise versus decline is rejected with over 99.99% confidence. Equal likelihood of negative and positive skewness on 5-year intervals is rejected with 99.8% confidence.

Simulations

For deeper analysis, we need to compare credit spreads to historical evidence on default risks. Figure 8.2 displays annual default rates since 1920 for corporates monitored by Moody's (Ou et al, 2024, Exhibit 36). While Moody's initially covered the US only, it has gradually expanded abroad. The number of issuers was about 3000 in the early 1920s, sank to around 600 in 1950, and has since risen to nearly 7000.

Figure 8.2 shows few signs of resets from a common distribution. Instead, ν seems to couple diffusion to nearby values with long-term mean reversion. If the process is Markov and the stationary distribution is gamma with shape D_0 and rate T_0, the diffusion must take the form

$$d\nu = \tfrac{1}{2} s^2 \nu^{2\tau-1}\left(D_0 + 2\tau - 1 - T_0 \nu\right)dt + s\nu^{\tau} dz \qquad (8.7)^{22}$$

for constant baseline speed s and power τ. This can approximate nearly any non-negative diffusion (Dennis and Patel 1984).

Figure 8.2: Annual US/global rated corporate default rates 1920-2023

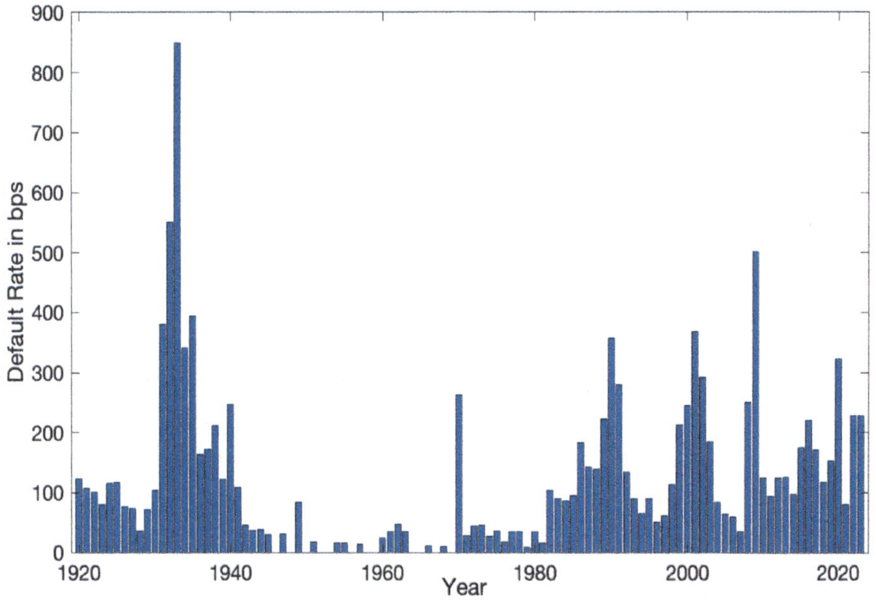

For higher ν to induce bigger moves, τ must be positive. The best-known application in finance is the model of Cox, Ingersoll and Ross (1985), where $\tau = \frac{1}{2}$:

$$d\nu = \tfrac{1}{2} s^2 (D_0 - T_0 \nu) dt + s\sqrt{\nu}\, dz \,. \tag{8.8}$$

The best-known application in evolutionary biology and population ecology is the May (1973) model of stochastic logistic growth, where $\tau = 1$. Applying Itô's Lemma, we can express it as

$$d \log \nu = \tfrac{1}{2} s^2 (D_0 - T_0 \nu) dt + s\, dz \,. \tag{8.9}$$

When D_0 is high, there is not much practical difference between $\tau = \frac{1}{2}$ and $\tau = 1$ as ν clusters near its mean E_0. When D_0 is low, as in Figure 8.2, default rates vary a lot, so τ significantly affects relative behavior at extremes. Regressions of yearly changes in default rates against their baseline values strongly favor τ close to 1.[23] Accordingly I chose (8.9) over (8.8) and other forms of (8.7).

Interestingly, $\log \nu$ in (8.9) behaves much like frailty Y in the Duffie (2011) model. The only structural difference is that the mean reversion is linear in ν rather than $\log \nu$. While substitution would make the stationary distribution for ν lognormal instead of gamma, the parameters can be tweaked to make their means and variances match.

I used the sample mean $E_0 \approx 120$ bps to estimate $T_0 = D_0 / E_0$. For a first-pass estimate of D_0, I divided E_0^2 by the sample variance to obtain 0.8. However, this likely understates D_0 as the yearly data smooths intra-year fluctuations in ν. For a first-pass estimate of s, I removed all years with zero defaults and took the square root of the mean squared log change to obtain 0.8. However, the removal likely understates s.

To refine the estimates, I converted the diffusion into a daily transition matrix Π.[24] For every (E_0, D_0, s) combination tested, I ran 10,000 simulations of (8.9) for the duration and pool sizes of the Moody's data and looked for relatively stable metrics to compare simulated averages with observed statistics. The best fits for $E_0 = 120$ bps rounding to tenths were $D_0 = 0.6$ and $s = 1.0$.[25]

Consider the following thought experiment. Imagine that

- Corporate ν diffuses for 100,000 years via (8.9) driven by the benchmark parameters.
- The ν induce random defaults in a pool of $N = 3000$ credits.
- A Perfect Learner predicts ν by processing defaults and servicing in (8.4) without any other information.

Figure 8.3 simulates 500 years of annual default risk. Ignore the scale on the right-hand side, which will be discussed in Chapter 10. Note the big variations from one 50-year period to the next. This underscores the difficulties of fitting (E_0, D_0, s) in (8.9) with less than two centuries of data. Our choices are no more than plausible proxies.

However, even a Perfect Learner who knows the correct parameters has trouble predicting ν, which frequently changes by more than 20% in a month. The high volatility encourages forecasters to give huge weight to recent news. Since N observations arrive at a time, our main measure of duration is T_{eff}/N rather than T_{eff} alone.

Figure 8.3: Simulation of annual default risk for benchmark for 500 years

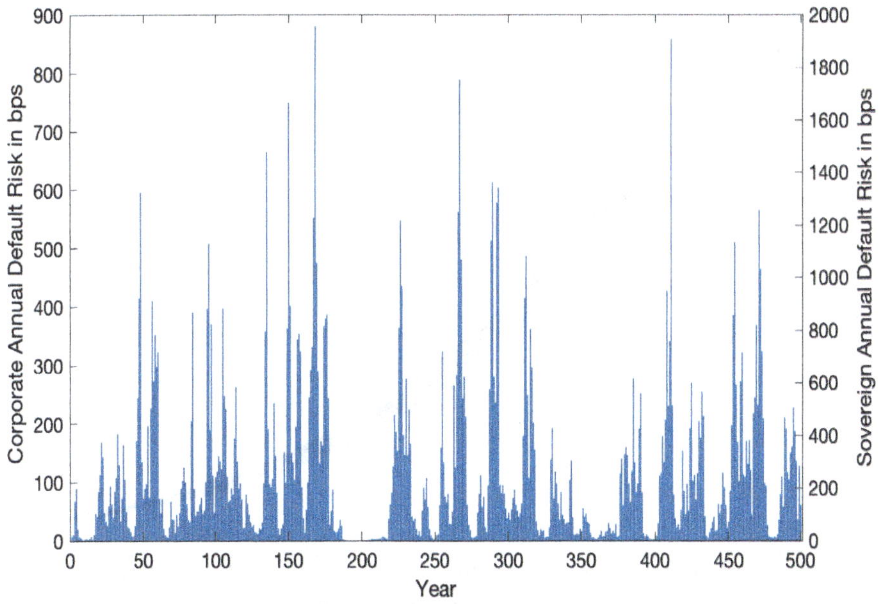

For a Perfect Learner, duration averages 3.5 months. While duration typically rises as E falls, as shown in Figure 8.4, it never reaches 15 months. The corresponding absolute jump $1/T_{eff}$ on news of an iid default ranges from less than 3 bps to more than 60 bps, with a mean of 17 bps. From a FIRE perspective, this looks irrationally myopic.

On average 3 defaults occur per month. D_{eff} averages 5 with 90% of values less than 10 and 10% less than 1. As Figure 8.5 shows, D_{eff} and E generally move together, with the safest debt the least certain. Halving E cuts D_{eff} by more than a third. For $E < 20$ bps, which occurs a quarter of the time, the mean jump on news of default is over half of E. In over half of those cases the mean jump on default exceeds E. Of course, defaults in those situations are rare.

Between low D_{eff} and low T_{eff}/N, the RMSE of daily E versus ν is 57 bps, nearly half of E_0. But that is not the Perfect Learner's fault. Between the high volatility of risk and the slow pace of defaults, the close tracking assumed by FIRE isn't close to being feasible. It makes it hard to know just how safe reputedly safe debt really is.

Figure 8.4: Scatter plot of T_{eff}/N versus E for corporate benchmark

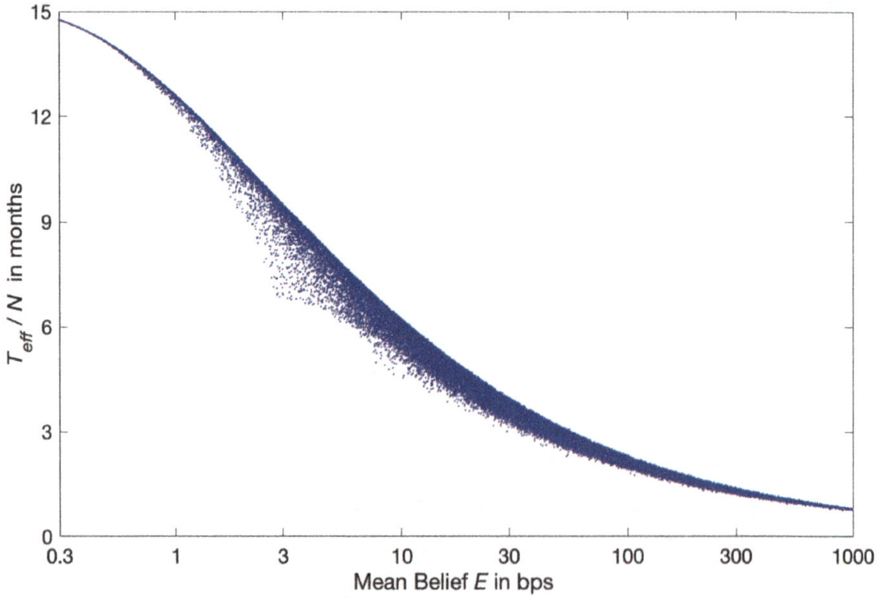

Figure 8.5: Scatter plot of D_{eff} versus E for benchmark

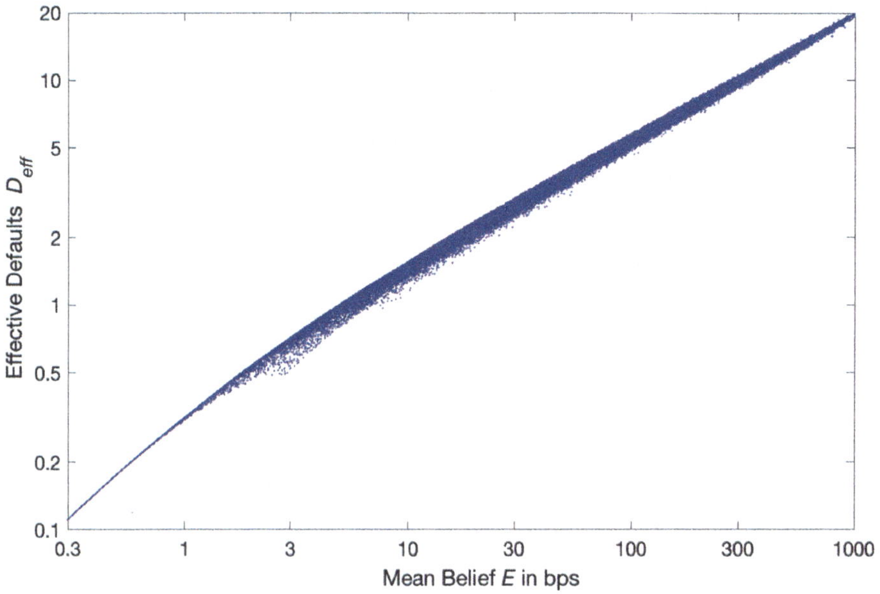

Bonds

For more insight, let's turn from immediate default risk to longer-term default risk. Let $S_t(\nu_j)$ denote the probability of servicing to year t for initial default risk ν_j. Then the probability $S_{t+dt}(\nu_i)$ of servicing dt longer equals $1-\nu_i dt$ times the transition-weighted average of the various $S_t(\nu_j)$ with vector form $S_t(\mathbf{v})$. For diagonal matrix diag($\mathbf{v}$), transition matrix Π, and $dt = 1/250$, recursion starting from the unit vector $S_0(\mathbf{v}) \equiv \mathbf{1}$ implies

$$S_T(\mathbf{v}) = \left(\Pi - \tfrac{1}{250}\mathrm{diag}(\mathbf{v})\Pi\right)^{250T}\mathbf{1}. \tag{8.10}$$

For a repayment T years in the future, the logarithmic default rate is $c_T(\nu) \equiv -\log S_T(\nu)/T$. For simplicity I assume zero salvage value on default, which makes c_T a fair credit spread absent risk premia and taxes. The fair credit spread given uncertainty is $C_T(p) \equiv -\log\langle S_T \rangle_p / T$. Credit spreads contract toward a long-term mean as T increases due to the mean reversion in ν.

I will focus on c_{10}, which corresponds to a "bullet" bond pledging a single payment 10 years forward. For our corporate benchmark, c_{10} ranges from 26 bps at $\nu = 1$ bp to 240 bps at $\nu = 1000$ bps. As c_T is concave in ν, $C_T(p)$ is less than $c_T(E)$, which in turn is less than the expected default rate $c_T^*(E)$. However, the differences $c_T^* - C_T$ for our benchmark rarely exceed 7 bps. Figure 8.6 charts c_{10}^* on top of a scatter plot for C_{10}. The scatter is quite thin for most E, reflecting the tight relation between D_{eff} and E. The spreads are slightly convex in $\log E$.

Suppose we tried to confirm empirically that a market applying C_{10} to the benchmark is pricing risk efficiently. The ideal dataset consists of 10-year bullets B_{nt} issued by firm n at time t at spread $C_{10}(t)$ for every n and daily t along with record of servicing 10 years later. Let us again posit $N = 3000$ firms with every defaulter issuing new bullets of equal risk. How many decades of data would we need for high confidence that prices forecast 10-year default risk well?

Figure 8.6: 10-year credit spreads for benchmark

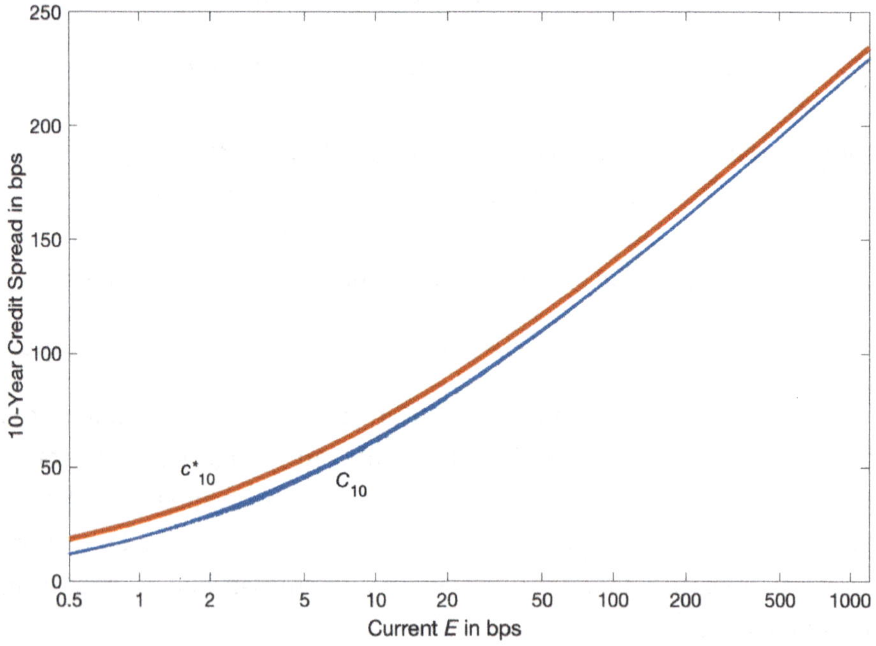

It is tempting to think that a few decades will suffice. The default risk for a 10-year bullet given average yearly risk of $E_0 = 120$ bps is 11.4%. For 3000 independent credits with the same risk, the standard deviation is 17.4 for the number of defaults or 58 bps for the default rate. The standard deviation drops to 18 bps for the average over 10 independent years and 11 bps for a 30-year average.

However, those estimates grossly understate the true variability. While the credits are conditionally independent, they share the same fluctuating risk ν. Look again at Figure 8.3, which charts sample yearly default risks for 3000 credits. Their standard deviation is 141 bps. While averaging ν over 10 randomly chosen years would shrink the standard deviation to 43 bps, neighboring years are highly correlated. The actual standard deviation of average ν for 10 consecutive years is 96 bps. In effect, one decade provides barely 2 independent yearly samples.

In a 100,000-year simulation, the annual default rate of our benchmark has a standard deviation of 93 bps. Given 10 consecutive 10-year samples with no overlap between them, the standard deviation drops to 38 bps, as the correlated decades provide effectively 6 independent yearly samples. If we examine overlapping 10-year samples, no matter how densely we overlap them, the standard deviation stays 38 bps. Actual defaults over 110 years have a 10% chance of differing by more than 50% from the number expected.

This makes it hard to check the predictive power of longer-term corporate debt. Checking the predictive power in different environments—e.g., C_{10} at low E versus C_{10} at high E—is even more challenging. While long-term bonds make long-term commitments, the market scales that weigh those commitments are always calibrated to current beliefs, as supply must balance demand. Few bond traders worry about long-term accuracy either. What matter far more are:

- identification of the consensus beliefs and short-term news that jointly drive current prices.
- anticipation of how the consensus is likely to persist or change.

The best we can hope for is a market that behaves like a Perfect Learner. Figure 8.7 displays Perfect Learner C_{10} and one-day E for a 100-year simulation. Since C_{10} spans a much narrower range than E, I graph them on separate scales to keep both paths clearly visible.

Bond spread is far calmer than bill spread. Since market price is $\exp(-C_T T)$, price volatility is T times the annualized standard deviation of ΔC_T, which I will call spread volatility. In a 100,000-year simulation, spread volatility for C_{10} averages 30 bps, which is barely one-quarter of mean $\langle C_{10} \rangle = 117$ bps and one-sixth of the mean 173 bps spread volatility for E.

For comparison, spread volatility on 10-to-15-year investment grade US corporate debt 1997-2024 also averaged 25% of the average spread (ICE BofA Data Indices 2024a). However, the latter is 171 bps, nearly 50% higher than for the benchmark. The next chapter will explore some causes.

Daily changes ΔC_{10} in the benchmark are highly skewed upward, with averages of 3.7 overall and 6.5 for C_{10} below the median. This reflects the low T_{eff} displayed in Figure 8.4. Observed 10-15-year spread changes are skewed similarly though not quite to that degree: overall average 2.8 and 4.3 for C_{10} below the median.

Figure 8.7: Simulated corporate credit spreads for 100 years

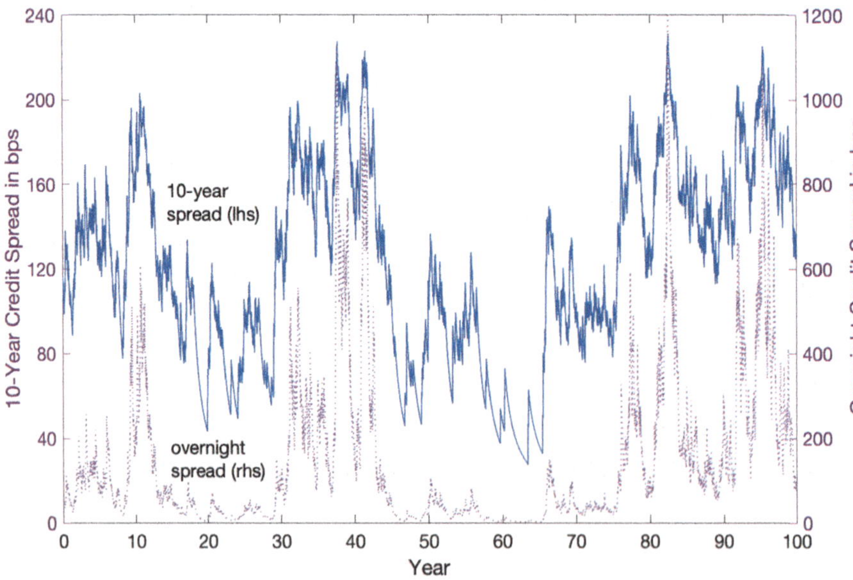

9

Credit Grades

As we have seen, defaults are too infrequent to estimate the risk without inference from a host of seemingly related assets. The assets are treated as loads of a common riskium prone to sporadic decay. There is enough corporate data to track average risk ν reasonably well even though ν is highly volatile. However, bond markets are not content to trade averages. They bet on the risks of specific credits. This catalyzed attempts to categorize firms by their relative levels of risk. These categories are known as credit grades and specialized firms have arisen to rate them.

By convention, most credit rating agencies assign at least seven distinct letter grades. They also subdivide major letter grades into three tiers, which are identified by numbers or signs. Table 9.1 displays average one-year global corporate default rates by credit grade for the two most comprehensive series I found. The data from Moody's (Ou et al 2024, Exhibit 26) cover 1920 through 2023. The data from S&P (Kraemer et al 2024, Table 24) cover 1981 through 2023.

Both lists exhibit roughly geometric spacing with over a thousand-fold differential across the whole. Each letter drop roughly triples the Moody's rate and quadruples the S&P rate. However, the Moody's gradient steepens for post-1980 data, where it is not significantly different from the S&P gradient. This suggests that rating agencies, not surprisingly, have learned over time to distinguish relative default risks better.

Table 9.1: Average annual global corporate default rates by letter credit grade in basis points

Moody's Rating	Default Rate 1920–2023	S&P Rating	Default Rate 1981–2023
Aaa	0	AAA	0
Aa	5	AA	1
A	9	A	5
Baa	26	BBB	18
Ba	102	BB	81
B	309	B	394
Caa-C	1027	CCC-D	2446

Figure 9.1 charts average default rates by tier on a log scale. The rates give equal weight to the longest samples available, which came from S&P for 1981-2023 and Moody's for 1983-2023.[26] The best log-linear fit accounts for 98.7% of the variance. Its slope of 0.48 per tier implies an average risk multiplier of $e^{3\times0.48} \approx 4.2$ per full-letter-drop in grade. The confidence interval spans 0.45 to 0.5 for a risk multiplier of 3.8 to 4.5.

Since the aggregate risk ν fluctuates considerably over time, each credit tier seems best viewed as an average multiplier ω applied to ν. Despite the steady gradient, capital markets place special emphasis on grades of Baa3/BBB- ("triple B" as I will call them) or higher. These are investment grades, which appeal to a broader class of lenders than the speculative grades of "double-B" or lower. Given average ν over the period of 155 bps, the investment-grade threshold of 39 bps corresponds to $\omega \approx 0.25$.

Historically, firms prized investment grade ratings and tended to withdraw from ratings (which they typically pay for) if they fell too far away. In recent decades, opportunities for speculative financing have greatly expanded, causing many higher-risk firms to seek ratings. Nowadays roughly half of ratings are speculative. Still, investment grades carry extra cachet and firms near the boundary strive hard to obtain them. Since 2000, both triple-B and single-B ratings have significantly outnumbered double-B.

Figure 9.1: Average default rates per tier for Moody's/S&P mix 1981-2023

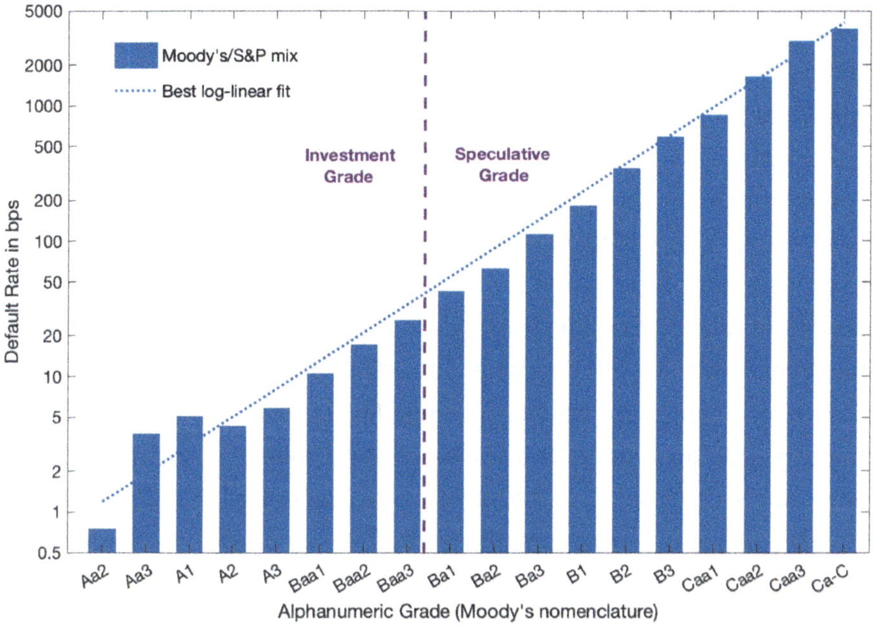

The main practical justification for special treatment of investment grades is that they hardly ever default at high rates. Since 1981, the maximum corporate default rate for the lowest triple-B tier was 194 bps for Moody's and 133 bps for S&P, compared to a maximum default rate for the highest double-B tier of 505 bps for Moody's and 370 bps for S&P.

Long-term investment-grade loans are riskier than these numbers suggest, since the loans' credit rating might decline before they mature. A common regulatory response distinguishes between lower investment grades where the risk of deterioration to speculative grade seems significant and higher investment grades where it does not. Banking regulators tend to demand small capital buffers (reserves against unexpected losses) for double-A or triple-A loans, moderate buffers for single-A loans, and significantly higher buffers for speculative-grade loans. While this approach has serious drawbacks, its influence increases the importance of estimating ω well and forecasting its likely evolution.

How do we estimate ω for a particular credit? Ideally, we find a sound indicator z of relative default risk, pool all credits with similar z, count the relevant D_{eff} defaults in T_{eff} years of relevant observations, and compare the default rate D_{eff}/T_{eff} to the aggregate $\langle \nu \rangle$. The best fitting ω will equal their ratio, with a relative standard deviation of $\sqrt{1/D_{eff}}$.

Suppose we aim to assign at least 90% of credits to the correct tier. For tiers not quite 0.5 log units wide, the relative standard deviation should be less than 25%, which requires $D_{eff} > 16$. Is there enough corporate data to achieve that? That depends on the tier. Assuming ω and z maintain a broadly stable relation for several decades, we have hundreds of defaults for every C and single-B tier (where over 90% of rated defaults occur) and dozens of defaults for double-BB tiers. For triple-B, Moody's reports 87 same-year defaults since 1981 (Ou et al 2024, Exhibit 47), so the lower tiers have ample coverage. But the same period generated only 19 defaults in Moody's single-A grades and 10 defaults for its double-AA grades. That makes it hard to distinguish internal tiers or the boundary with triple-B. The corresponding defaults for S&P are even fewer: 73 for triple-B and 10 for all A-grades (Ou et al 2024, Exhibit 47).

As noted in Chapter 8, good identification hinges on finding a statistical "riskium" relationship between z and ω that spans investment and speculative grades. Suppose $\omega = z\mathrm{X}$ for lognormal noise X, while analysts estimate α and β in $\omega \approx \alpha z^{\beta}$ drawing on a few decades of data on defaults with $\langle \nu \rangle \approx 120$ bps and with z distributed in proportion to Moody's/S&P corporate ratings. While the relative RMSE in estimation increases near-linearly with log distance from single-B, it is usually less than 0.2 (40% of the width of a tier) even for triple-A. However, hardly any simulations generate defaults in triple-A or higher double-A tiers, making it hard to confirm any salient differences between them.

Market prices suggest general confidence that low credit risks can be inferred from high credit risks. Figure 9.2 plots daily credit spreads since 1997 for indices of the 8 main grades of US corporates (ICE BofA Data Indices 2024a). The logarithmic scale shows a relatively stable gap between them despite big fluctuations in a seemingly common driver, consistent with our modeling of risk as ω times aggregate ν.

Figure 9.2: Credit spreads for US corporate bonds by major credit grade

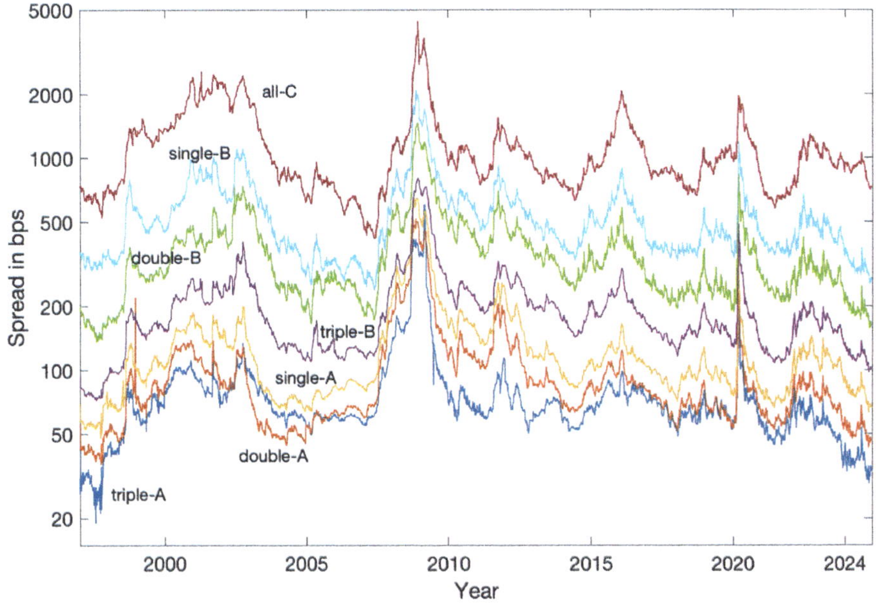

The principal component of spread correlations is nearly perfectly correlated with the average spread and hence can be associated with ν. Its variance accounts for 90% of total spread variance. Its daily changes account for over 50% of the total variance of daily changes. However, the tracking is weaker toward the edges.[27] Changes in the principal component are about 80% correlated with spread changes in single-A through single-B but only 66% correlated with C-grade changes and 60% with triple-A changes. Volatility of log changes averages 26% for all grades except double-A at 38% and triple-A at 43%.

Figure 9.3 plots the ratios of credit spreads to the triple-B spread. Higher systemic risk related to the financial crisis of 2008/9 compressed the relative differentials between credit grades. Note also the instability of the double-A to triple-A ratio. This seems to reflect a mix of weak discernment between the two grades and dwindling numbers of triple-A credits. As credit markets get more sophisticated, with more willingness to fund speculative-grade risks, the advantages of super-safe reputation are receding. The main remaining edge for triple-A is lower risk of downgrade.

Figure 9.3: Ratios of credit spreads to triple-B spreads

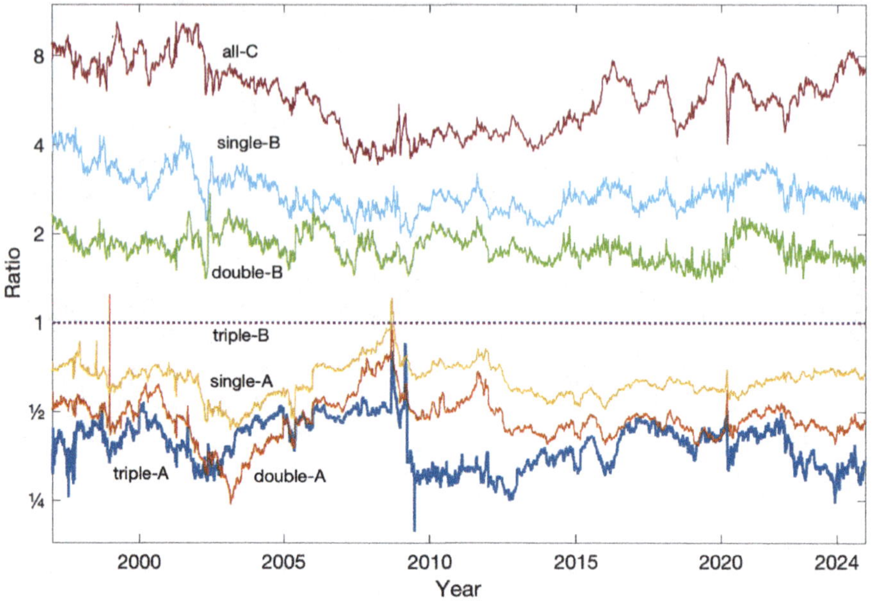

Credit Migration

Let us next analyze migration across credit ratings. Table 9.2 displays Moody's average annual transition rates by letter grade for 1920 to 2017 (Ou et al 2024, Exhibit 31). Defaults are categorized separately as are credits that withdraw from ratings. The table can be regarded as a transition matrix P, whose elements P_{ij} approximate the probability that a credit starting in category i ends the year in category j. Here are a few features shared by every ratings transition matrix:

- Migrations imply significant compression of forward default rates.
- Migration rates typically rise as credit quality declines.
- Withdrawal rates rise as credit quality declines.
- Downgrades are slightly more common than upgrades except in the lowest tiers.
- Outer tails are fatter than Gaussian, which suggests some combination of measurement errors and jumps.

Table 9.2: Average 1-year transition rates in percent for corporate credits 1920-2023 from Moody's (DF=default, WD=withdrawn)

To: From	Aaa	Aa	A	Baa	Ba	B	Caa	Ca-C	DF	WD
Aaa	87.17	7.47	0.77	0.18	0.03	0.00	0.00	0.00	0.00	4.38
Aa	0.98	84.65	7.51	0.68	0.15	0.04	0.01	0.00	0.06	5.92
A	0.06	2.57	86.01	5.14	0.56	0.10	0.03	0.01	0.08	5.46
Baa	0.03	0.20	3.84	84.30	4.02	0.63	0.11	0.02	0.23	6.63
Ba	0.01	0.06	0.44	5.97	74.73	6.61	0.67	0.08	1.10	10.32
B	0.00	0.04	0.14	0.55	5.44	72.34	6.32	0.44	2.99	11.74
Caa	0.00	0.01	0.02	0.08	0.35	5.86	70.34	2.94	6.69	13.71
Ca-C	0.00	0.01	0.07	0.07	0.38	2.35	9.83	44.85	25.60	16.83

For more detailed analysis I will use Moody's average migration rates across 20 tiers since 1983 (Out et al 2024, Exhibit 24). The first step converts the larger matrix P into a smaller matrix Π that conditions on no exit through withdrawal or default. One method assigns all exits to the tier at the start of the year. That is, if we partition P into $\left[P_{trunc} \quad P_{DF} \quad P_{WD}\right]$, Π equals P_{trunc} with values for $P_{DF}+P_{WD}$ added to the diagonal. Another method, favored by Elton et al (2001), assigns them in proportion to tier at the end of the year, by rescaling the rows of P_{trunc} to sum to one. As tests showed little difference, I used their average.[28]

Absent exit, average credit rating declines at a rate of about 2 tiers per decade in double/triple-A, 1 tier per decade in single-A, and 0.5 tiers per decade in B-grades. Only the bottom Ca-C tier rebounds sharply, with upgrade expected within 3 years absent default or withdrawal in the interim. However, since ω rises exponentially with tier, the expected changes in tier significantly understate the expected changes in risk. Expected default rates grow on the order of 50% per year in double/triple-A, 30% per year in single-A, and 25% per year in double/triple-B. Expected rates then decline roughly linearly from 21% in B1 to 4% in Caa2. Only C-Ca, down 11%, can expect a big reduction in conditional risk.

For high-grade credits, the expected increases $\langle \Delta\omega \rangle \langle \nu \rangle$ in default risk given are tiny: 0.1 bp for triple-A, 0.7 bps for double-A, and 1-2 bps for single-A using the data from Figure 9.1. These rise to 3-9 bps for triple-B, 17-28 bps for double-B and 44-74 bps for single-B. The estimates for C-grade vary wildly, reflecting their instability: +93 bps for Caa3, +64 bps from Caa2, -21 bps for Caa1, and -400 bps for Ca-C.

Assuming a stable Markov process, the implied increases in expected default risk 10 years forward are 5 bps for triple-A, 14 bps for double-A, 22-39 bps for single-A, 55-125 bps for triple-B, 200-330 bps for double-B, and 420-490 bps for single-B and Caa1. That switches to reductions of 580 bps for Caa3 and 1900 bps for Ca-C. To rephrase the shifts as ratings, looking 10 years forward triple-A risk is expected to behave like A2 risk and double-A risk like Baa1 risk. Single-A through double-B risks are expected to drop 3 tiers while single-B risks are expected to drop 2 tiers. Absent default, all C-grades are expected on average to achieve Caa1 risks.

Hence, portfolios of long-term corporate bonds should expect both major compression of relative hazard rates (conditional default risks) over time and major rises in average hazard rates. To say this crudely, most long-term bond portfolios slide toward single-B. How can we reconcile this with the persistently broad spectrum of ratings? The core explanation is differential survival rates. The market of rated credits continually purges weaker credits through defaults and withdrawals from ratings.

Figure 9.4 displays the average life of credits in different tiers, which is computed $(\mathrm{I}-\mathrm{P}_{trunc})^{-1}\mathbf{1}$ for identity matrix I. Credits in A-grades stay rated without default for an average 18-20 years. The corresponding lives of speculative credits decline near-linearly from 12 years for Ba1 to 2.7 years for Ca-C. The new credits that enter the ratings pool are much stronger on average than the credits that leave.

Figure 9.4: Average corporate rating life by tier for Moody's transitions 1983-2023

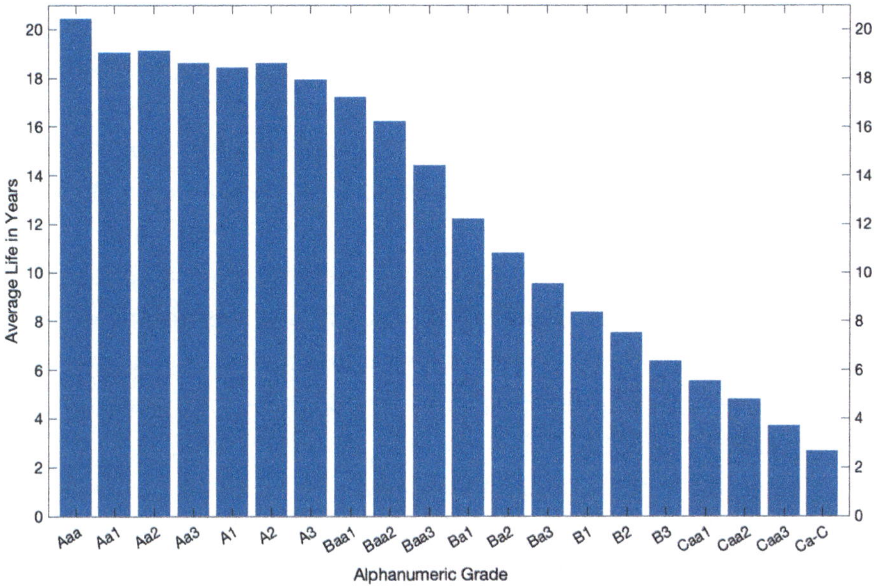

Between long rating lives and good prospects for keeping default risks less than 100 bps, investment grade credits mostly merit their reputation as relatively safe, predictable long-term investments. This appeals to insurance companies and other investors obliged by either regulation or temperament to maintain high buffers against long-term losses. The evolution of speculative-grade risks is much harder to predict, partly due to their high multiplier ω on unstable aggregate ν and partly due to the high variance of their changes $\Delta\omega$. Since corporates generally have better information on their $\Delta\omega$ prospects than lenders do, a C-grade corporate seeking a long-term loan raises doubts about its viability. Accordingly, speculative-grade debt rarely has long maturities.

For illustration, the giant investment manager BlackRock offers corporate bond funds in various categories. Its biggest investment-grade fund, with a market cap of $30 billion, has a weighted average maturity of 13 years. In contrast, its biggest high-yield fund, with a market cap of $19 billion, has a weighted average maturity of barely 4 years.[29]

Credit Risk Premia

The most perplexing aspect of Figure 9.2 is its scale. In investment grades, the mean spreads exceed the default rates in Table 9.1 by 75-150 bps. In speculative grades the nominal gaps exceed 300 bps once we factor in expected salvage values. The relative markups on default risks look huge, especially for double/triple-A debt which is nearly risk-free short-term.

Some of the markup reflects expected credit deterioration. While the anticipated rate of deterioration is slow for high-grade bonds, their maturities tend to be long. Annual single-digit basis point risk often grows to double digits while double-digit risk often grows to triple digits. However, even after adjustment for deterioration, default risk covers less than 10% of A-grade credit spreads and less than half of B-grade spreads (Huang and Huang 2012).

Taxes add another wedge that is relatively constant across grades. Interest is taxed on US corporate bonds but not on US Treasury bonds. As a result, even permanently risk-free corporate bonds need to pay more interest than Treasuries to cover their tax burden. Elton et al (2001) and Liu et al (2007) attributed most A-grade spreads to taxes.

Corporate bonds also get discounted for their relative illiquidity compared to Treasuries. Liu and Wu (2004), Longstaff et al (2005), Bao et al (2011) and De Jong and Driessen (2012) all report significant impact but differ on its relative importance. Definitions of illiquidity vary, while fluctuations in estimated levels often are so correlated with changes in aggregate risk ν that it is difficult to distinguish the two effects.

Between these factors and secular variations in ν, it is hard to identify the residual risk premia for default risks. Nozawa (2017) focused on cross-sectional variations in spreads. Assuming that tax and liquidity effects are relatively stable, spread changes must reflect fluctuations in either expected credit losses or discount rates. Nozawa found that expected credit losses explained firm-level changes much better than risk premia but that the two factors had roughly equal influence on aggregate market spreads. This implies that average risk premia are substantial although it does not elucidate their value.

Credit default swaps (CDS) potentially allow good identification since they bet directly on default. Comparing 1.2 million pairs of CDS rates and Moody's predictions of corresponding default risks, Berndt et al (2018) found that risk premia averaged 2-3 times expected losses with huge variations over time. They also found that risk premia increased on average by about half as fast as expected losses.

A similar relationship is evident in Figure 9.2. Credit spreads roughly double with each full-letter drop in grade while default rates roughly quadruple. Hence the Berndt et al findings circle back to my initial observations, which I previously qualified as misleading. Judging from their Figure 4, Berndt et al use term structures of default probabilities that are far less steep than we inferred from Moody's data. Indeed, they posit large reductions two years forward in the risks of double-B, single-B and triple-C credits, which the transition matrix in Table 9.2 cannot support.

Perhaps the correlation between equities and speculative-grade debt can account for the risk premia on the latter. As regards investment-grade debt, the default risks are so low that even a dozen bps in risk premium would entice massive purchases from half-to-full Kelly gamblers were it not for the low net return on principal. If the gamblers could leverage their exposures cheaply enough, say by borrowing 97% of the bond price at a risk-free rate from a third party, investment would be worthwhile. But in that case both the third-party lender and investment-grade borrower would be better off contracting directly.

While the investment-grade bond market is massive, most holders are insurance companies, pension funds and banks (Koijen and Yogo 2023). Holders typically have big liabilities they aim to secure, with regulators pressing them to do so far more through concentration in safer assets than through diversification across riskier assets. Volumes are further swelled by favorable tax treatment of debt issuance, especially the deductibility of interest paid.

In short, the investment-grade debt market seems quite segmented from equity markets (Goldberg 2015, Jung et al 2020). Since credit spreads can be viewed as insurance fees, I searched for insights from actuaries, who have long puzzled over appropriate profit margins. A comprehensive review by Rubin et al (2009) identified seven different actuary approaches to "quantifying margins for uncertainty". None of them offer an explicit formula rooted in default risk and risk aversion. The only formula I found came from the CDS literature. Unfortunately, its key multiplier for risk aversion is the ratio of observed prices to estimated risks. The empirical tweak helps fit the data but does not justify its size.

While I have not found a solution, here is where I suspect it lies. What do institutional holders of investment-grade debt worry about most? Not average defaults or small fluctuations around the average, as those are expected. Not gradual increases in risks either, as maturing lower-yielding bonds will be replaced with higher-yielding bonds. What institutional holders most dread is news of dramatically worsening risk, which devalues even bonds that don't default. The longer the duration of holdings, the greater the drop in value and the longer the likely wait until full recovery. If too many liabilities come due too soon, say through a run on demand deposits or withdrawal of repo-ed securities, the holder might fold.

The risk that much of the portfolio goes down together cannot be reduced to correlated default risk, any more than icebergs can be reduced to correlated ice (Osband 2001). In statistical terms, they involve high-order cumulants, where the triangles of exchangeable probability from Chapter 1 extend to ever-wider rows. In finance terms, big clustering risks hinge on a surging common factor, like the ν we have described.

Banks are highly vulnerable to clustering risks since they are highly leveraged with big duration mismatch: their assets generally mature much later than their liabilities. This is euphemistically called "maturity transformation" and justified as a promoter of investment. However, it comes at considerable cost: aggravation of boom-and-bust cycles, extended slumps when banks are forced to tighten, wasteful lending when losses are bailed out, and erosion of public trust.

These tensions have spawned extensive regulation to ensure that banking is safe, largely by making banks identify risks, monitor them and ensure that they stay within permitted bounds. While the resulting risk management is increasingly sophisticated, it is inherently torn between fulfilling the spirit of the regulations, fulfilling the letter of the regulations without regard to spirit, and exploiting grey areas and loopholes in the regulations. Details about the risks of individual assets or peculiar portfolios often obscure the views of the main common risks.

Kotlikoff (2010) and Cochrane (2014) recommend measures to make banks run-proof, simplify enforcement, reduce regulatory burden, and arguably improve the financial system overall. The most important element is the drastic reduction in duration risk. King (2016), a former Governor of the Bank of England, contends that the illusory "alchemy" transforming risky, illiquid assets into safe money-like claims is the core problem with the global financial system.

Duration entails much higher tail risks than most regulators expect. Consider a portfolio of 3000 equally weighted credits. If default risk stays constant at 120 bps, the annualized standard deviation of the default rate will be 20 bps. In 99.9% of years, excess loss rates on a portfolio of 1-year bullets will be less than 64 bps. Yet for the benchmark process for 10-year bullets modeled in Chapter 8, the prices on 10-year bullets have a standard deviation of 275 bps with a 99.9-percentile drawdown of 930 bps. Factoring in credit deterioration and fluctuations in risk premia would push the drawdown well past 1000 bps.

I call attention to the 99.9 percentile because a common rule of thumb deems capital buffers as safe if they are projected to cover 100% of risks 99.9% of the time. By this criterion, a bank holding only the benchmark portfolio described above cannot safely leverage its reserves ten-fold. Since standard measures generally understate the fluctuations in common risks, and since even a few decades without crisis inflates confidence that risks are nil, 99.9-percentiles are generally set too low.

Worse, leverage that seems perfectly justified short-term can be perilous later. Suppose regulators significantly lower the required buffers on loans to a particularly safe sector. Bank lending soars, which raises asset prices, makes loans even safer, and encourages even more lending to that class. If the news is too good for too long, a bubble will form and pop. As valuations decline, banks will be forced to sell off portfolios and restrict further lending, exacerbating the decline and potentially causing a larger financial bust.

The regulatory reforms known as Basel II (Basel 2004) provide a case study. They allowed banks extra leverage for reputedly safe investments. Long-dated housing mortgages were especially favored as no major country-wide waves of defaults had occurred in the previous fifty years. The resulting surge of mortgage lending encouraged rapid appreciation of housing, which in turn lowered lending standards on prospects of housing collateral repaying the mortgage regardless of buyer income. Once the boom started to recede, defaults soared, which triggered sharp cutbacks in lending, plunging prices of housing, and a global financial crisis.

The crisis encouraged tighter control by central banks with broader guarantees of bailouts, but again with little attention to duration mismatch. A recent example is the failure of Silicon Valley Bank, which invested most of its short-term liabilities in long-term US bonds and collapsed when interest rates rose. Remarkably, the bank was deemed systemically important enough to warrant depositor bailout yet had been allowed to run a huge duration mismatch.

10

Sovereign Risk

The main conclusion of Reinhart and Rogoff's banner study of sovereign debt crises, highlighted in its title *This Time is Different*, is that sovereign debt markets are prone to manic mood swings. When things go well for an extended period, lenders tend to underestimate risks. They are too short-sighted even to anticipate their own vulnerability to self-fulfilling panic:

> Perhaps more than anything else, failure to recognize the precariousness and fickleness of confidence [...] is the key factor that gives rise to the this-time-is-different syndrome. Highly indebted governments, banks, or corporations can seem to be merrily rolling along for an extended period, when *bang!*—confidence collapses, lenders disappear and a crisis hits. (Reinhart and Rogoff 2009, xxxiv-xxxix)

This chapter shows that rational myopia can account for most of these phenomena. Rational myopia can be useful since it speeds recovery after crisis. However, our analysis supports Reinhart and Rogoff's warning (2009, xxv) that "excessive debt accumulation [...] often poses greater system risks than it seems during a boom". Indeed, it implicitly calls for extra vigilance by policymakers to check what lenders almost surely won't.

For historical data on sovereign debt, I draw mainly on spreadsheets maintained online by Reinhart through 2012, supplemented with Moody's data for 2013 to 2023 (Liu et al 2024). Reinhart and Rogoff's core measures identify the entire interval between default and resolution as the time in default. To convert these into default rates I treat the first year of each default episode as the time of default and the rest of the episode as a limbo not registering either default or servicing. Each year's default rate equals the number of default episodes starting that year divided by the number of issuers not in default the previous year.

Figure 10.1 depicts annual sovereign default rates from 1850 through 2023. Their average is approximately 270 bps, over twice the average corporate rate in Figure 8.2. The annual rates jump around far more than corporate rates do, mostly because the pools are much smaller: 40 countries in 1850, 60 in 1945, and 125 in 2012.

I fit the mean-reverting diffusion (8.8) to sovereign ν applying the same methods that I used for corporates. The best fits given $E_0 = 270$ bps for D_0 and s are close enough to the corporate fits $D_0 = 0.6$ and $s = 1.0$ that I will use the same values. This treats the sovereign benchmark analogously to a high-yield corporate benchmark, in line with the findings of Gopinath et al (2025) that the two markets behave similarly.

Figure 10.1: Sovereign default rates on external debt 1850-2023

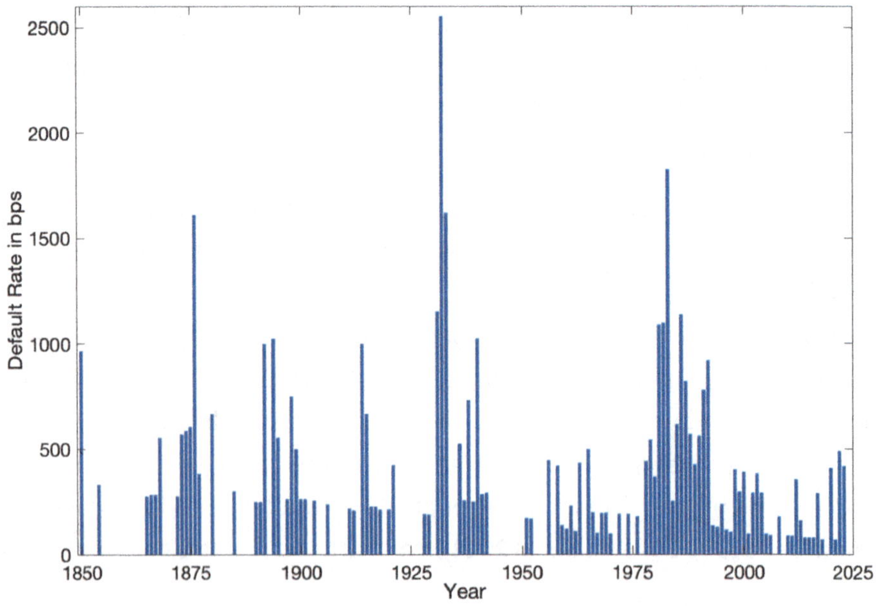

It follows that simulations of corporate default risks can be viewed as simulations of sovereign default risks when rescaled by the factor $270/120 = 2.25$. This explains the right-hand axis of Figure 8.3. Given the huge variations over 50-year spans and the relatively small pool sizes of sovereigns, our confidence in any parameterization is rightly low.

To keep the modeling simple, the next two charts posit a Perfect Learner monitoring a pool of $N = 100$ sovereigns. Figure 10.2 is a scatter plot of duration T_{eff}/N versus E for a 100,000-year simulation. Mean duration is 9.6 months versus 3.5 months for the corporate benchmark; 99%. of values fall between 5 months and 17 months. The mean jump on news of default is 139 bps, eight times the mean corporate jump. The minimum jump is 72 bps while the maximum jump exceeds 250 bps.

The ratio $1/D_{eff}$ of jump to baseline E tends to be very high for sovereigns. Only 4% of D_{eff} exceed 4, while 70% are less than 2 and 37% are less than 1. Like for corporates, the scatter plot in Figure 10.3 of sovereign D_{eff} versus E is near-linear. D_{eff} is less than 1 for all $E < 110$ bps,

Figure 10.2: Scatter plot of T_{eff}/N versus E for sovereign benchmark

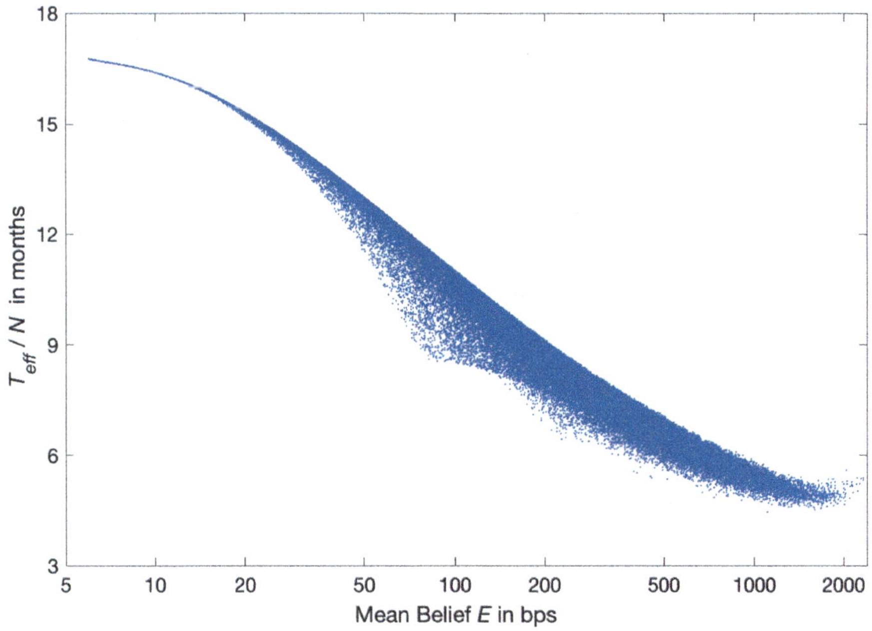

Figure 10.3: Scatter plot of D_{eff} versus E for sovereign benchmark

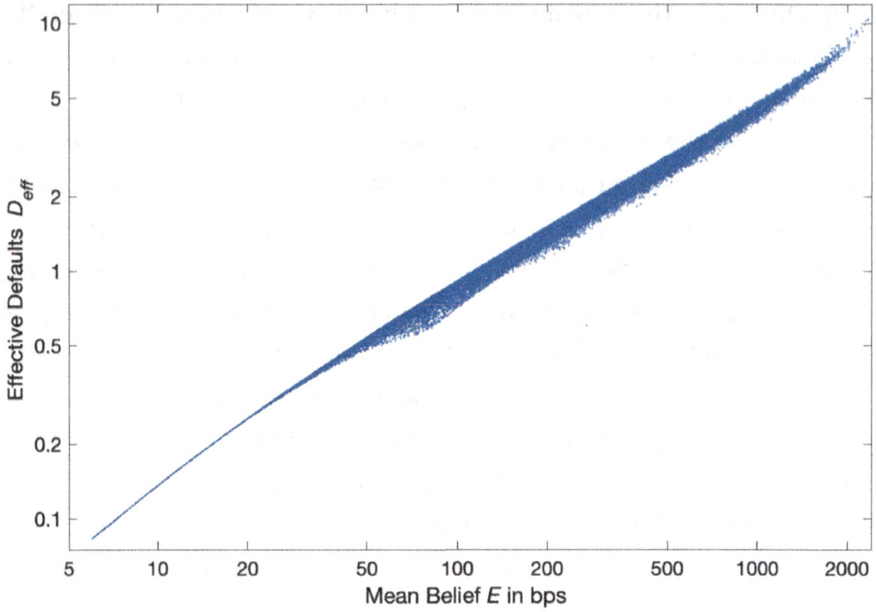

less than 2 for all $E < 290$ bps, and less than 2.5 for all $E < 395$ bps. Over 40% of defaults occur in this region, and when they do the high relative jumps will make the risks appear to have been greatly undervalued. Hence Perfect Learners will seem unusually prone to understate sovereign risks, in line with Reinhart and Rogoff's observations, albeit without a trace of irrationality.

The predicted jumps in spreads for 10-year bullets are relatively modest. They average 32 bps for a 320 bps drop in price. However, our benchmark calculations ignore the huge uncertainties about the true drivers of ν or the appropriate credit-grade multiplier ω. News can amplify their impact.

According to Reinhart and Rogoff, lenders tend to significantly understate sovereign risks on the eve of default. To some extent this is inevitable. For any given E, defaults are more likely to occur for high ν than low ν, so conditioning on the eve of default overweights $\nu > E$ relative to $\nu < E$. Similarly, it overweights $E > C_{10}$ relative to $E < C_{10}$.

The differentials are much wider for the sovereign benchmark than for the corporate benchmark. For corporates on the eve of default, average ν of 289 bps is 21 bps higher than average E and 129 bps higher than average C_{10}.. For sovereigns on the eve of default, average ν of 712 bps is 174 bps higher than average E and 379 bps higher than average C_{10}. Hence, Reinhart and Rogoff are justified in highlighting the differentials though not in inferring the systematic understatement of sovereign risks. Indeed, recent work by Meyer, Reinhart and Trebesch (2022) indicates that sovereign lending tends to be profitable.

Might taking a long view be worthwhile to offset complacency on good news or panic on bad? Consider the following thought experiment. Imagine that the market uses an EMA of 100 years duration (EMA_{100}) to make all credit forecasts. Figure 10.3 displays a 1000-year simulation with mean close to the long-term norms. Absolute yearly changes average 5 bps and only 1% exceed 25 bps. Furthermore, the market never gets extremely pessimistic. In a 100,000-year simulation, only 1% of EMA_{100} values exceed

Figure 10.4: Simulation of EMA with 100-year duration for sovereign risk

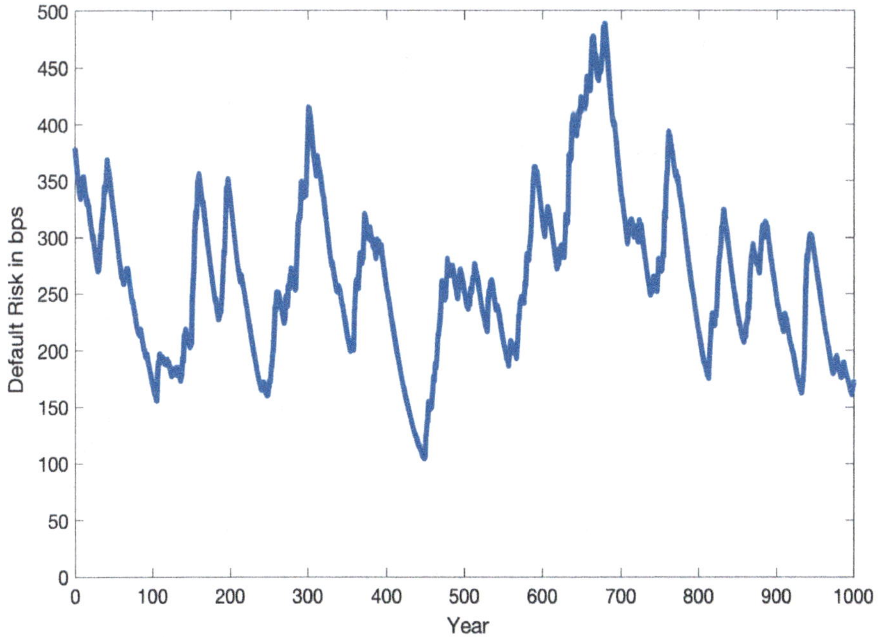

480 bps compared to a 17% chance of Perfect Learner $E > 480$ bps and a 2.5% chance of Perfect Learner $E > 1000$ bps.

The flip side is that it is hard to get confident that risks are low. Although Perfect Learner E is 33% likely to be less than 100 bps, EMA_{100} dips below 100 bps less than 2% of the time. Restoration of confidence is glacial. EMA_{100} needs 35 years to halve its value even when there are no defaults in the interim. In contrast, two years without default suffice to halve any Perfect Learner E greater than 150 bps.

Hence, EMA_{100} ties credit spreads to decades-old behavior, regardless of improvements since. Overall tracking is poor too. The median absolute deviation from v is 180 bps for EMA_{100} versus 74 bps for Perfect Learner E.

Rollover Risk

Short-term debt is often deemed information-insensitive except in crisis, when fears of default suddenly soar (Dang et al 2020) This is a loose descriptor for the learning process described in (8.4), where servicing induces slow decline towards a positive floor and news of relevant default triggers a large discrete jump. True insensitivity to information implies long periods where short-term credit spreads do not change. This is rare.

Granted, structural information about GDP growth, fiscal discipline, or governance can significantly affect future rates in ways that T-bill holders hardly care about. While our model can proxy that by changing the risk multiplier ω, it is hard to gauge the appropriate adjustment. Sovereigns vary far more in governance, economy, accounting systems, culture, environment, and key risks than US corporates and their pool is relatively small. Since it is harder to find comparable countries, lenders naturally place extra weight on the sovereign's own servicing history.

For example, Hungary and Poland are geographically close, were both members of the Soviet bloc with similar economic and political systems and had similarly high foreign debt burdens of around 60% of GDP in 1989 after the Soviet bloc collapsed. However, their debt servicing strategies were sharply different. Hungary was adamant to service while Poland defaulted. Not surprisingly, Poland's bonds paid much higher interest rates than Hungary's in the 1990s, although Poland's growth rates were higher and its finances less constrained.

Servicing news also bears directly on future rollover risk, a major structural concern. Most sovereign debt is refinanced through issuance of new debt. The longer the maturity, the less frequent the rollovers, but there is always some chance that debt is refinanced under duress, with extra servicing costs that worsen fiscal burdens and make default more likely. The degree of risk depends less on current debt trends than on public confidence in their management. The more reliable the past servicing, even if financed purely by rollovers, the greater the confidence it will continue, which in turn facilitates further rollover.

The best way to maintain that confidence long-term is to occasionally wind down the relative debt burden. As Hoffman et al (2007) have noted. Britain did this repeatedly during the 18th century, thanks largely to lender sway in Parliament. Good reputation allowed Britain during the Napoleonic Wars to accumulate debt exceeding 250% of GDP without triggering a crisis. By the eve of the first World War, nearly a century later, public debt had shrunk to around 25% of GDP without default, devaluation, or major surges in credit spreads.

In contrast, French kings in the 18th century paid at least 200 bps higher interest rates than their English rivals and frequently defaulted, as they paid little heed to lender confidence. In 1788 Louis XVI convened a long-dormant assembly of elites to secure approval of higher taxes. Instead, the assembly insisted on constitutional reforms and hamstrung tax collection. The ensuing revolution financed itself through paper money *assignats* that lost 99% of their value in five years. This ruined lenders and crippled long-term French credit markets until the 1850s.

While rollover can defer fiscal reckoning, it cannot permanently deny it. Having limited authority to impose domestic taxes, the kings of late medieval Spain financed their wars through loans repaid out of silver from imperial mines in Mexico and Peru. Between mine exhaustion and ships lost to storms and pirates, revenues disappointed. Ten times between 1557 and 1662, Spain suspended payments and forced creditors to take long-term *juros* bonds as salvage value. Eventually the *juros* defaulted too and wrecked Spain's financial markets (Hoffman et al 2007).

Confidence in rollover is a network externality that reduces short-term default risk. Each lender is reassured by the prospect that some other lender will pay the future interest and principal. That is especially important for fiat money. This peculiar form of sovereign debt has no maturity and earns no interest but can be used to pay taxes and settle other debts. Between the state-bestowed legitimacy, the ease of handling, and the lack of intrinsic value, fiat money is a useful means of purchase. But why should sellers of intrinsically valuable goods and services accept it? The main reason is confidence that others will accept it too, thanks to the host of monetary payments that occur every day. The pool size is enormous, so confidence is easily developed and maintained.

Doubts about fiat money focus on the degree of inflation undermining money's real worth. While central banks often set inflation targets, these are better viewed as mean-reverting influences than reliable predictions. Accordingly, inflation expectations largely project from past inflation trends. Given rational learning under uncertainty, these expectations should (i) rely heavily on simple error correction, (ii) behave much more myopically than fully informed expectations, and (iii) behave least myopically (often described as "well anchored") when inflation trends seem stable. Hajdini (2023) confirms (i) and (ii), while Moessner and Takáts (2020) confirm (iii).

Might it be possible to create money that has no intrinsic value, broad acceptability, and strict supply rules that are insulated from government fiscal policies? That is the promise of cryptocurrency and in principle it seems feasible. However, the leading cryptocurrencies are currently far too volatile to provide even roughly stable purchasing power. Their main current attractions are the speculative bubbles they entail and their concealment from taxation and reporting requirements. Stablecoins, which are cryptocurrencies intended to be bubble-free, have been vulnerable to value-destroying runs and fraud. I do not know how the engineering challenges will be mastered.

Rational Bubbles

Imagine a major sovereign borrower that is very wealthy, has a sterling servicing record, and manages a fiat currency with global acceptability, but persistently fails to scale back its mounting debt? Should we

- project its default risk from a host of poorer, weaker countries with worse servicing records and shakier reputations?
- treat its fiat currency debt as risk-free thanks to its option to print any money it needs?
- infer long-term risks from the spreads on long-term bonds?

Unfortunately, none of these approaches are trustworthy. Recall the difficulty in estimating low-ω corporate default risk even with thousands of defaults in higher tiers. Sovereigns offer at most a few hundred relevant defaults for study. While debt issuance in one's own currency facilitates servicing, so does rollover at higher interest rates, and neither of these rule out default. As for credit spreads, we have previously noted that long-term prediction is not a capital market forte and that incentives encourage a short-term focus.

This section highlights an additional complication, drawing on Osband et al (2024). Even when fiscal policy is clearly unsustainable, with default ultimately certain, short-term debt markets can stay calm for a decade or more, with spreads that give few hints of crisis. In the interim the debt inflates and becomes a rational bubble.

To make the logic clear, I will start with the simplest model that embeds it. Imagine a market economy maintaining perfect stability in real output and prices. The budget is perfectly balanced except for previously accumulated debt. The government does not want to default, as that would disrupt the payment system and eliminate credit lines that might be needed later. Yet it refuses to run the fiscal surplus needed to extinguish the debt, as transferring real resources to creditors could reduce perceived domestic welfare. Instead, the government issues overnight T-bills at the market-determined rate and aims to keep rolling over the debt indefinitely without tapping own resources.

Creditors are risk-neutral and indifferent to delay; they hold T-bills as long as the expected overnight returns remain non-negative. If creditors were certain of repayment, T-bills could sell at par. However, the transactions are tinged with fear. The government worries that creditors might use T-bill repayments to purchase goods and services, sparking inflation and reducing other consumers' welfare. Creditors worry that the government might default. These fears reinforce each other. The more creditors worry about default, the more tempted they become to switch to goods. The more the government fears a creditor switch, the more it considers preemptive default.

Let us assume a fearful equilibrium where lenders always correctly perceive the current hazard rate $h(t)$. The greater the T-bill stock $B(t)$, the greater the risk of default since more real resources might be drained upon redemption. The simplest tractable formulation makes h an increasing power function of B. Dropping the t argument as understood,

$$h = \kappa B^{\alpha} . \tag{10.1}$$

Perceived risk forces the T-bills to offer a credit spread c over the presumed risk-free interest rate of zero. Assuming all maturing T-bills are rolled over into new T-bills until default occurs,

$$B' = cB = \kappa B^{\alpha+1} . \tag{10.2}$$

Assuming no salvage value on default, $c = h$ in equilibrium. Subscripting the initial values with 0, the only feasible solution is

$$B(t) = B_0 \left(1 - \alpha c_0 t\right)^{-1/\alpha} \tag{10.3}$$

For large values of t, the implied $B(t)$ is negative. On inspection, B blows up to infinity in a finite time $\Omega = 1/(c_0 \alpha)$. I will call Ω the Debt Apocalypse because default is certain by then. We can confirm this by examining the hazard rate. Since $c = (\log B)'$,

$$h(t) = c(t) = \frac{c_0}{1 - \alpha c_0 t} = \frac{1}{\alpha(\Omega - t)} . \tag{10.4}$$

As $t \to \Omega$, h outstrips any finite bound, ensuring default before Ω.

In mathematical terms, B is a local martingale: a fair game from one moment to the next that might not be fair long term. In effect, each lender focuses only on short-term payoffs while presuming that some others will lend longer-term. Osband et al (2024) discusses this in more depth and provides links to related research on financial bubbles.

What is most intriguing about (10.4) is the long deferral of crisis. From Figure 9.1, each full letter drop in credit grade roughly quadruples the default risk. By that metric, $\alpha \approx 2$ if doubling B knocks one letter off the credit grade, while $\alpha \approx 3$ if doubling B knocks 1.5 letters off the credit grade. If α lies in that range, then for any $c_0 < 50$ bps it takes decades for c to double. Risks do not mount rapidly until t gets close to Ω.

While Osband et al (2024) adduce some evidence for $2 < \alpha < 3$, the relevant α might be even higher. As we have seen, past servicing tends to keep rollover risks low, so rising B might initially have little impact on c. However, once c gets large enough to warn of significant risk, bonds lose their appeal as safe havens and risk-averse holders look to limit their exposure. To attract speculative lenders, c must rise further, which causes a political dilemma. If the government pays, speculators earn much more than expected, which strikes the broader public as unfair. If the government continually reissues at high c, it implicitly admits that it is desperate for financing and that fears are well-founded, which makes both sides prone to panic. If B is a large share of GDP, few sovereigns will pay $c > 500$ bps for more than a few years before defaulting.

Let us now generalize the model for more realism:

- Default risk depends not on nominal debt B by itself but on the ratio b of B to GDP Y. This converts (10.1) into

$$h = kb^{\alpha} . \tag{10.5}$$

- GDP grows at constant relative rate g:

$$Y' = gY . \tag{10.6}$$

- Creditors demand a risk-free return r, which bumps up the interest rate i to r plus the credit spread c:

$$i = r + c . \tag{10.7}$$

- The primary budget surplus equals a constant fraction q of the bond stock B. While not the standard way to depict budget trends, it is the simplest to analyze, as it lets us replace (10.2) with

$$B' = iB - qB. \qquad (10.8)$$

- If default occurs, creditors lose a fraction η of the T-bills' face value. The expected rate of return is $r + c - \eta h$, so in equilibrium

$$c = \eta h. \qquad (10.9)$$

From (10.5) and (10.9), $(\log c)' = \alpha(\log b)'$. Substitution of (10.6)-(10.8) into $(\log b)' = (\log B)' - (\log Y)'$ indicates $(\log b)' = c + \delta$ for $\delta \equiv r - g - q$. Hence $c' = \alpha c(c + \delta)$, which can be rewritten as

$$\frac{dt}{dc} = \frac{1}{\alpha c(c+\delta)} = \frac{1}{\alpha\delta}\left(\frac{1}{c} - \frac{1}{c+\delta}\right).$$

Integrate both sides, set $t = 0$ at spread c_0, and rearrange to obtain

$$c(t) = \frac{c_0 \delta e^{\alpha\delta t}}{\delta + c_0 - c_0 e^{\alpha\delta t}}. \qquad (10.10)$$

For $c_0 + \delta > 0$, this applies only until a Debt Apocalypse at time

$$\Omega = \frac{\log(1 + \delta/c_0)}{\alpha\delta}. \qquad (10.11)$$

Most economists seem confident, along the lines of Blanchard (2019), that any debt stock is sustainable provided $i < g$ and that low c_0 testifies to market confidence in sustainability. Equations (10.10) and (10.11) challenge this view. The core sustainability metric is $c_0 + \delta < 0$ or $g > i - q$.

Figure 10.5 depicts the evolution of $c(t)$ for various δ given benchmark values $\alpha = 2.5$ and $c_0 = 50$ bps. For $|\delta| < 2\%$, most changes are modest for two decades or more. Expected corrections in fiscal policy further flatten the curve. Osband et al (2024) show that market uncertainty can easily obscure the risk of major crisis.

Ideally, we should distinguish good debt expansion from bad debt expansion. Good debt expansion seeks only to smooth the taxes associated with emergency expenditures (Barro 1979). Bad expansion invokes emergencies on flimsy pretexts to delay needed fiscal tightening.

Figure 10.5: Evolution of credit spreads given $\alpha = 2.5$ and $c_0 = 50$ bps

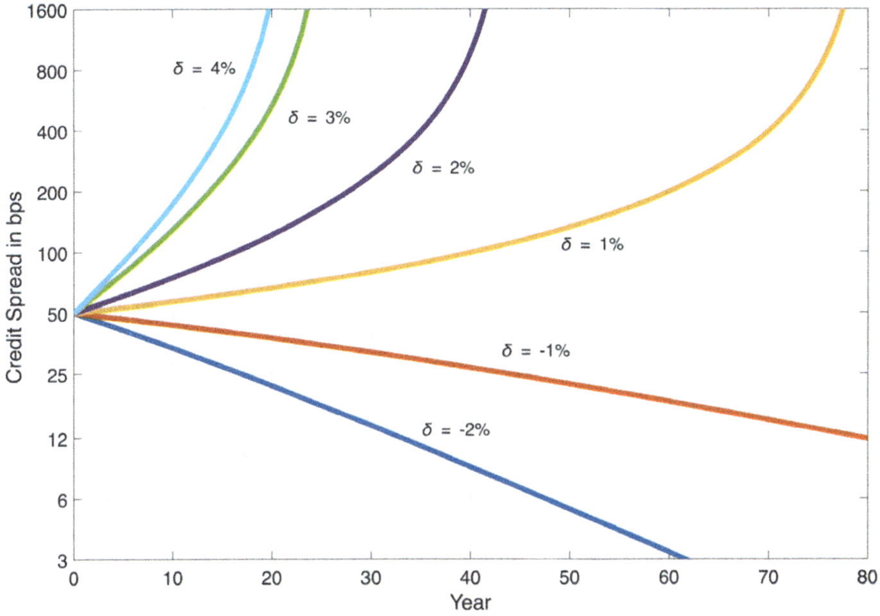

Figure 10.6 illustrates the challenges. It depicts US government debt held by the public from 1939 to 2022. Here the public includes the Fed. The dotted line excludes Fed holdings where available.[30]

US debt soared during World War II despite cheap financing. In 1947 and 1948, the gross debt-to-GDP ratio fell by 27 percentage points, partly because the primary fiscal balance flipped into surplus but mostly because inflation soared after wartime price controls were removed. Over the next quarter century, δ averaged -4% annually, mostly due to $r-g$.

In contrast, δ has averaged over 3% since 1980 and nearly 5% since 2000, reflecting higher $r-g$ and mostly negative q. Resistance to fiscal tightening is strong. A rapidly greying population keeps expanding entitlements for seniors. Rising inequality encourages more aid to lower tiers. Higher taxes threaten to crimp economic growth. Most fiscal debates pit lower taxes against higher benefits; in practice each side wins some of what it wants, which lets both spending and deficits escalate.

Figure 10.6: Publicly held US debt/GDP on a log scale

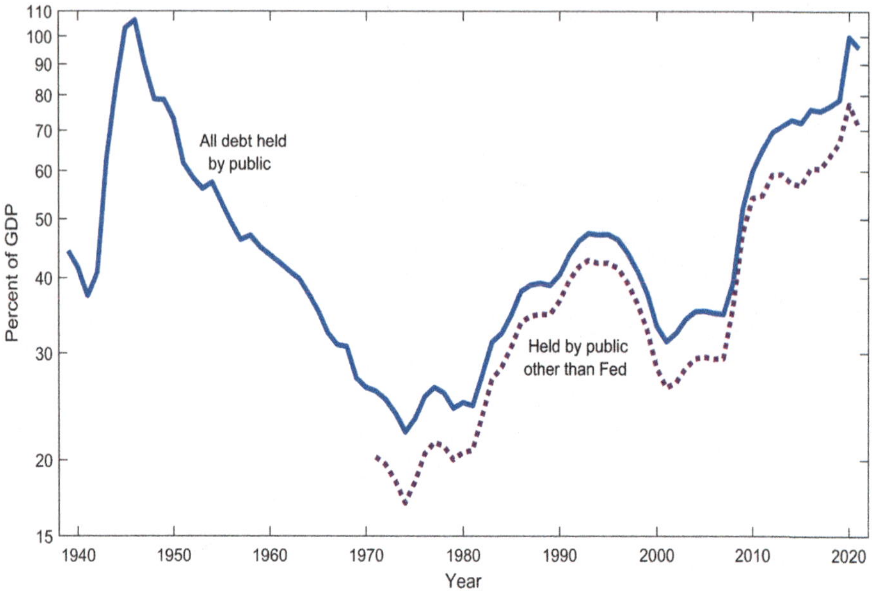

Growing partisanship coupled with a nearly even electoral split leads each side to begrudge any concessions to the other (Yared 2019). When bond markets tremble, the Fed often steps in with bond purchases and other measures to reduce volatility. Many politicians have inferred that debt and deficits are irrelevant.

In short, the US has grown addicted to high government debt. With less pressure to make hard tradeoffs, too many groups grab too much. The next generation adds to the previous bill and consigns it to the generation after. Obligations to repay morph into perceived rights to keep borrowing.

RATE doesn't directly address these questions. It just warns against the FIRE treatment of credit spreads as reliable answers. A debt burden that continually mounts as a share of GDP is like tinder piling up in a forest. Bond markets act sometimes like deer grazing in the clearings and sometimes like forest rangers scanning for fires. Rarely do they react to the tinder before it alights. Still, the tinder matters, as it can turn a brush fire into a raging inferno.

Notes

[1] Turbulent flow in fluids is characterized by swirling vortices. Vortices reflect conservation of angular momentum. Any analogues in finance would require multidimensional analysis beyond the scope of this book.

[2] This concept is much better appreciated in machine learning than it is in economics. For deeper treatments, see Frongillo et al (2015) and Casalaina-Martin et al (2017).

[3] One violation I do want to acknowledge involves aversion to vaguely defined *p*, which Aydogan et al (2023) call ambiguity aversion. This might influence the risk premium for high-grade corporate bonds, a puzzle I discuss in Chapter 9.

[4] See en.wikipedia.org/wiki/ Rabbit–duck_illusion. The image dates from an 1892 German magazine.

[5] While often attributed to Keynes, the original source appears to be Read (1898): "It is better to be vaguely right than exactly wrong."

[6] EVSI is not the only information measure we could use. The K-L divergence compares entire distributions and not just their means. I compared EVSI and expected K-L divergence for 10 million plays of a Perfect Learner. For $(\alpha,\beta,\lambda)=(1,1,0.005)$ the correlation was 0.914. I could not drive the correlation below 0.8 without making λ much smaller or moving the long-term mean well away from 0.5.

[7] To generate the same result using normality, note that

$$p(\Delta x|\mu) \propto \exp\left(-\frac{(\Delta x-\mu\Delta t)^2}{2\sigma^2\Delta t}\right) \propto \exp\left(\frac{\mu}{\sigma^2}\Delta x-\frac{\mu^2}{2\sigma^2}\Delta t\right) \approx 1+\frac{\mu}{\sigma^2}\Delta x,$$

where the last step follows from a second-order Taylor expansion and the heuristic approximation $(\Delta x)^2 \approx \sigma^2\Delta t$. For the updated $p+\Delta p$ to integrate to one, $1+(\mu/\sigma^2)\Delta x$ must be divided by $1+(\kappa_1/\sigma^2)\Delta x$ or multiplied by about $1-(\kappa_1/\sigma^2)(\Delta x-\kappa_1\Delta t)$. Taking the limit as $\Delta t \to 0$ yields (4.2).

[8] Merton (1980) emphasized this to encourage more attention to volatility estimation in high-frequency financial data.

[9] Let me emphasize that K^{shift} is not a function of the Markov parameters alone. It depends on p and therefore constantly varies.

[10] Any single Kalman filter works much better for objects that change course gradually than for objects that dart about. Arrays of Kalman filters can provide more flexible tracking, with credibility shifting from one filter to another, but those shifts can be turbulent.

[11] Barsky and De Long (1993) made the same core argument: "An estimated long-run growth rate [of dividends] is highly correlated with the level of dividends. Prices therefore respond more than proportionately to long-run movements in dividends.". However, they did not relate the correlations to the variance of beliefs or frame learning as wholly rational.

[12] Given a diffusion in p with drift $\mu(p)$ and volatility $\sigma(p)$, the probability density $f(p,t)$ satisfies the Fokker-Planck equation

$$\frac{\partial f}{\partial t} = -\frac{\partial(\mu f)}{\partial p} + \tfrac{1}{2}\frac{\partial^2(\sigma^2 f)}{\partial p^2}.$$

For any p bounded at zero, the stationary f^* must be proportional to

$$\frac{1}{\sigma^2(p)}\exp\left(\int_0^p \frac{2\mu(w)}{\sigma^2(w)}dw\right).$$

Here $\mu(p) = \lambda(\pi_H - p)$ and $\sigma(p) = p(1-p)S_{MAX}$.

[13] The chosen metric was the average absolute log deviation from empirical averages for mean yearly growth, standard deviation, reset rate, probability in crisis, and mean drawdown in crisis. For every parameter combination examined, means were proxied by the sample averages in a 10-million-year simulation. I used Matlab's *fminsearch* to minimize.

[14] In a multi-layer market it can be hard to precisely identify $\mathrm{cov}(\bar{Q},\mu)$ in (4.7), which in turn makes it hard to identify ς in (4.8). The estimator $\Delta\log Q/\Delta y$ is usually very good but can be terrible if $|\Delta y|$ is small to $\langle Q^{shift}\rangle$. For robustness I used the square root of the ratio of two EMAs: one for $(\Delta\log Q)^2$ and the other for $(\Delta y)^2$, I set EMA duration at 3 days each.

[15] My simulations assume driftless Brownian motion priced at F tracked by EMAs of learning rates η with variance $V^* = \eta/(2-\eta)$ around 0. They repeatedly confirm that a Kelly fraction $F^2/(F^2+V^*)$ is nearly

optimal. However, mean log growth plummets when $(0-F)^2$ is replaced by $(E-F)^2$, as this overweights the most extreme values of E.

[16] In simulations, the ratio of mean absolute deviation to standard deviation is typically least when $V_{traders}$ is low as that makes tiny doubts more significant.

[17] That cannot be wholly correct. Applying (4.11) to calculate $\langle dQ \rangle$, $\langle dP \rangle$ includes an additional $\frac{1}{2}\bar{Q}''(F)dt$ times $(E_i - F)^2 + V_i - V$, where V_i denotes the variance of trader i's beliefs. This term averages to zero over all traders and is generally small. However, it favors some extra betting by traders with outlier beliefs.

[18] The first two chapters of *On the Origin of Species* (Darwin 1859) are entitled "Variation under Domestication" and "Variation under Nature".

[19] Martin (2013) showed that Epstein-Zin preferences preserve the risk premium (7.4) but change (7.3) to $Q^{-1} = \rho - \xi \mathrm{K}(1-\gamma)$ for $\xi \equiv (\psi^{-1}-1)/(\gamma-1)$. Given $\gamma > 1$, $\psi > 1$ would make Q decrease implausibly with μ. Yet Nakamura et al (2013) needed $\psi > 1$ to bring their fitted γ down to 6.

[20] Julliard and Ghosh (2012) draw the same conclusion from study of consumption patterns and excess returns.

[21] This follows from (i) the constraint $\nu \geq 0$, (ii) the Bayesian update factor $e^{-\nu t}$ for time t without default driving exponential decay in the tails and (iii) the tendency for p to peak near either 0 or the perceived E. A mixture of gamma distributions provides a better fit and helps explain changes in D_{eff} and T_{eff}.

[22] From note 12, for a diffusion in non-negative ν with drift $\kappa_1(\nu)$ and variance rate $\kappa_2(\nu)$, the stationary distribution f^* must be proportional to $\frac{1}{\kappa_2(\nu)}\exp\left(\int_0^{\nu} \frac{2\kappa_1(w)}{\kappa_2(w)} dw\right)$. For f^* to be gamma-distributed, κ_2 must be a power function of ν while κ_1/κ_2 must be affine in ν^{-1}. This implies (8.7).

[23] For yearly default rates v and next-year changes Δv, regressions of $|\Delta v|$ against v^{τ} fit v far better than $\sqrt{v}$. When both were included the coefficient on $\sqrt{v}$ was negative. The best fit was $\tau = 1.17$ although the gain over $\tau = 1$ was slight: adjusted R^2 rose by less than 1%. Since noise scales near-linearly with v, a regression of $|\Delta v|/v$ against v and a constant

should be more reliable apart from needing to drop the 12 observations with $v=0$. The coefficient on v was -0.1 and not statistically significant.

[24] I used a grid of 300 points that spanned 99.99% of feasible ν. Starting with $dt=1/250$ to represent one trading day, I halved it M times in succession until all transitions were confined to immediate neighbors and then squared the matrix M times to restore a daily interval.

[25] I tossed the roughly 90% of simulations generating average default rates more than 5% away from E_0. I then looked for metrics that were relatively stable for a given (D_0,s) pair but differed enough across different pairs to identify a good combination. I chose (i) the standard deviation of the default rates and (ii) the mean absolute change in yearly default rates.

[26] S&P does not report decompositions of C-grade debt, while Moody's phased in its C-grade decompositions between 1987 and 1998. The default rates for Caa1 and Caa3 have averaged respectively 0.5 and 1.8 times the corresponding average reported rate for Caa2. I imputed missing Moody's data by assuming these ratios held throughout and by assigning the aggregate Caa rate to missing Caa2. I imputed the C-grade counterparts for the S&P as 1.4 times the corresponding Moody's values, in line with the average ratios for medium or low single-B. I then averaged the yearly values with the issuer-weighted values reported for the period as a whole.

[27] When default rates in one grade are regressed against default rates in other grades, only the rates in adjoining grades are highly significant.

[28] The only noticeable differences involve the lowest Ca-C and Ca1 tiers, whose forward default rates rise respectively 2% and 1% per year faster when exits are allocated according to end-year grade.

[29] Data comes from www.ishares.com on 3 January 2025. The two funds are the iBoxx $ Investment Grade and the Broad US High Yield.

[30] Data comes from fred.stlouisfed.org, specifically Federal Debt Held by the Public and Federal Debt Held by Federal Reserve Banks.

References

Andersen, Torben G. 1996. Return Volatility and Trading Volume: An Information Flow Interpretation of Stochastic Volatility. *Journal of Finance* 51(1), 169-204.

Andersen, Torben G. and Luca Benzoni. 2009. Realized Volatility. In T.G. Andersen, R.A. Davis, J.-P. Kreiss and T. Mikosch, eds., *Handbook of Financial Time Series*, 555-575. Springer-Verlag.

Andersen, Torben G. and Tim Bollerslev. 1997. Heterogeneous Information Arrivals and Return Volatility Dynamics: Uncovering the Long-Run in High Frequency Returns. *Journal of Finance* 52(3), 975-1005.

Anscombe, Frank J. and Robert J. Aumann. 1963. A Definition of Subjective Probability. *Annals of Mathematical Statistics* 34(1), 199–205.

Asness, Clifford S., Tobias J. Moskowitz, and Lasse Heje Pedersen. 2013. Value and Momentum Everywhere. *Journal of Finance* 68(3), 929-985.

Aydogan, Ilke, Loïc Berger, and Valentina Bosetti. 2023. Unraveling Ambiguity Aversion. *Review of Economics and Statistics*, 1-32.

Baker, Rose D. and Ian G. McHale. 2013. Optimal Betting Under Parameter Uncertainty: Improving the Kelly Criterion. *Decision Analysis* 10(3), 189–199.

Bakshi, Gurdi and Georgios Skoulakis. 2010. Do Subjective Expectations Explain Asset Pricing Puzzles? *Journal of Financial Economics* 98(3), 463-477.

Bansal, Ravi and Amir Yaron. 2004. Risks for the Long Run: A Potential Resolution of Asset Pricing Puzzles. *Journal of Finance* 59(4), 1481-1509.

Bao, Jack, Jun Pan and Jiang Wang, 2011. The Illiquidity of Corporate Bonds. *Journal of Finance* 66(3), 911–946.

Barberis, Nicholas, Andrei Shleifer, and Robert Vishny. 1998. A Model of Investor Sentiment. *Journal of Financial Economics* 49(3), 307-343.

Barro, Robert. J. 1979. On the Determination of the Public Debt. *Journal of Political Economy* 87(5), 940–971.

Barro, Robert J. 2006. Rare Disasters and Asset Markets in the Twentieth Century. *Quarterly Journal of Economics* 121(3), 823–866.

Barro, Robert J. and José F. Ursúa. 2012. Rare Macroeconomic Disasters. *Annual Review of Economics* 4(1), 83–109.

Barsky, Robert B. and J. Bradford De Long. 1993. What Does the Stock Market Fluctuate? *Quarterly Journal of Economics* 108(2), 291-311.

Basel Committee on Banking Supervision. 2004. *International Convergence of Capital Measurement and Capital Standards. A Revised Framework.* Bank for International Settlements.

Bekaert, Geert and Guojun Wu. 2000. Asymmetric Volatility and Risk in Equity Markets. *Review of Financial Studies* 13(1), 1–42.

Benter, William, 1994. Computer Based Horse Race Handicapping and Wagering Systems: A Report. Chapter 19 in Donald B. Hausch, Victor Sy Lo and Williman T. Ziemba, *Efficiency of Racetrack Betting Markets* 2008, 183-198. World Scientific Publishing Co.

Berndt, Antje, Rohan Douglas, Darrell Duffie, and Mark Ferguson. 2018. Corporate Credit Risk Premia. *Review of Finance* 22(2), 419-454.

Black, Duncan. 1948. On the Rationale of Group Decision-making. *Journal of Political Economy* 56(1), 23–34.

Black, Fisher. 1976. Studies of Stock Price Volatility Changes. *Proceedings of the Business and Economics Section of the American Statistical Association*, 177-181.

Blanchard, Oliver. 2019. Public Debt and Low Interest Rates. *American Economic Review* 109(4), 1197–1229.

Bollerslev, Tim and Dan Jubinski. 1999. Equity Trading Volume and Volatility: Latent Information Arrivals and Common Long-Run Dependencies. *Journal of Business and Economic Statistics* 17(1), 9-21.

Bollerslev, Tim, George Tauchen and Hao Zhou. 2009. Expected Stock Returns and Variance Risk Premia. *Review of Financial Studies* 22(11), 4463-4492.

Bollerslev, Tim, Michael Gibson, and Hao Zhou. 2011. Dynamic Estimation of Volatility Risk Premia and Investor Risk Aversion from Option-Implied and Realized Volatilities. *Journal of Econometrics* 160(1), 235–245.

Born, Benjamin, Zeon Enders, and Gernot J. Müller. 2023. On FIRE, News, and Expectations. Working Papers of the Priority Programme 1859. *Experience and Expectation. Historical Foundations of Economic Behavior*. Humboldt University.

Brav, Alon, John R. Graham, Campbell R. Harvey, and Roni Michaely. 2005. Payout Policy in the 21st Century. *Journal of Financial Economics* 77(3), 483-527.

Breiman, Leo. 1996. Bagging Predictors. *Machine Learning* 24, 123–140.

Carhart, Mark M. 1997. On Persistence in Mutual Fund Performance. *Journal of Finance* 52(1), 57–82.

Carr, Peter and Liuren Wu. 2006. A Tale of Two Indices. *Journal of Derivatives* 13(3), 13–29.

Carr, Peter and Liuren Wu. 2009. Variance Risk Premiums. *Review of Financial Studies* 22(3), 1311–1341.

Casalaina-Martin, Sebastian, Rafael Frongillo, Tom Morgan, and Bo Waggoner. 2017. Multi-Observation Elicitation. *Proceedings of Machine Learning Research* 65, 1-27.

Chakravorti, Tatiana, Vaibhav Singh, Sarah Rajtmajer, Michael McLaughlin, Robert Fraleigh, Christopher Griffin, Anthony Kwasnica, David Pennock, and C. Lee Giles. 2023. Artificial Prediction Markets Present a Novel Opportunity for Human-AI Collaboration. *Proceedings of the 22nd International Conference on Autonomous Agents.*

Chellel, Kit. 2018. The Gambler Who Cracked the Horse-Racing Code. *Bloomberg Businessweek*, May 3.

Chen, Long, Zhi Da, and Richard Priestley. 2012. Dividend Smoothing and Predictability. *Management Science* 58(10), 1834-1853.

Chen, Yiling and David M. Pennock. 2010. Designing Markets for Prediction. *AI Magazine* 31, 42-52.

Chetty, Raj. 2006. A New Method of Estimating Risk Aversion. *American Economic Review* 96(5), 1821-34.

Chiappori, Pierre-André and Monica Paiella. 2011. Relative Risk Aversion is Constant: Evidence from Panel Data. *Journal of the European Economic Association* 9(6), 1021–52.

Chin, William and Marc Ingenoso. 2007. Risk Formulae for Proportional Betting. In Stewart N. Ethier and William R. Eadington, eds., *Optimal Play: Mathematical Studies of Games and Gambling*, 541-550. Institute for the Study of Gambling, University of Nevada.

Christoffersen, Peter, Kris Jacobs, and Chayawat Ornthanalai. 2010. Dynamic Jump Intensities and Risk Premiums: Evidence from S&P 500 Returns and Options. *Journal of Financial Economics* 106(3), 447-473.

Cochrane, John H. 1994. Permanent and Transitory Components of GNP and Stock Prices. *Quarterly Journal of Economics* 109(1), 241-266.

Cochrane, John. H. 2014. Toward a Run-free Financial System. Chapter 10 in Martin N. Bally and John B. Taylor, eds., *Across the Great Divide: New Perspectives on the Financial Crisis*, 197-249. Hoover Institution Press.

Cont, Rama. 2001. Empirical Properties of Asset Returns: Stylized Facts and Statistical Issues. *Quantitative Finance* 1(2), 223–236.

Cox, John C., Jonathan E. Ingersoll and Stehen A. Ross. 1985. The Term Structure of Interest Rates. *Econometrica* 53(2), 519–543.

Dang, Tri Vi, Gary B. Gorton, and Bengt R. Holmström. 2020. The Information View of Financial Crises. *Annual Review of Financial Economics* 12, 39-65.

Daniel, Kent, David Hirshleifer, and Avanidhar Subrahmanyam. 1998. Investor Psychology and Security Market Under- and Overreactions. *Journal of Finance* 53(6), 1839-1885.

Darwin, Charles. 1859. *On the Origin of Species by Means of Natural Selection, or the Preservation of Favoured Races in the Struggle for Life.* John Murray.

David, Alexander. 1997. Fluctuating Confidence in Stock Markets: Implications for Returns and Volatility. *Journal of Financial and Quantitative Analysis* 32(4), 427–462.

de Finetti, Bruno. 1931. Probabilism: A Critical Essay on the Theory of Probability and on the Value of Science. Translated in *Erkenntnis* 31, 239–261.

de Finetti, Bruno. 1937. La prévision: Ses lois logiques, ses sources subjectives. *Annales Institute Henri Poincaré* 7, 1–68. Translated in H. E. Kyburg Jr. and H. E. Smokler eds., *Studies in Subjective Probability.* 1980. Robert E. Kreiger Publishing Co.

de Finetti, Bruno. 1979. *Theory of Probability: A Critical Introductory Treatment.* Wiley.

De Jong, Frank and Joost Driessen. 2012. Liquidity Risk Premia in Corporate Bond Markets. *Quarterly Journal of Finance* 2(2), 1250006.

De Long, J. Bradford, Andrei Shleifer, Larry H. Summers, and Robert J. Waldmann. 1990. Noise Trader Risk in Financial Markets. *Journal of Political Economy* 98(4), 703-738.

Dennis, Brian and Ganapati P. Patel. 1984. The Gamma Distributions and Weighted Multimodal Gamma Distributions as Models of Population Abundance. *Mathematical Biosciences* 68, 187-212.

Der Kiureghian, Armen and Ove Ditlevesen. 2009. Aleatory or Epistemic? Does It Matter? *Structural Safety* 31(2), 105-112.

Ding, Zhuanxin, Clive W.J. Granger and Robert F. Engle. 1993. A Long Memory Property of Stock market Returns and a New Model. *Journal of Empirical Finance* 1(1), 83-106.

Drechsler, Itamar and Amir Yaron. 2011. What's Vol Got to Do with It. *Review of Financial Studies* 24(1), 1–45.

Duffie, Darrell. 2011. *Measuring Corporate Default Risk.* Oxford University Press.

Dumas, Bernard, Alexander Kurshev, and Raman Uppal. 2009. Equilibrium Portfolio Strategies in the Presence of Sentiment Risk and Excess Volatility. *Journal of Finance* 64(2), 579–629.

Ellsberg, Daniel. 1961. Risk, Ambiguity, and the Savage Axioms. *Quarterly Journal of Economics* 75(4), 643-669.

Elsemüller, Lasse, Martin Schnuerch, Paul-Christian Bürkner, and Stefan T. Radev. 2024. A Deep Learning Method for Comparing Bayesian Hierarchical Models. *Psychological Methods*.

Elton, Edwin J., Martin J. Gruber, Deepak Agrawal and Christopher Mann. 2001. Explaining the Rate Spread on Corporate Bonds. *Journal of Finance* 56(1), 247–277.

Engle, Robert F. 1982. Autoregressive Conditional Heteroskedasticity with Estimates of Variance of UK Inflation. *Econometrica* 50(4), 987–1008.

Engle, Robert F. 2001. GARCH 101: Use of ARCH/GARCH Models in Applied Econometrics. *Journal of Economic Perspectives* 15(4), 157-168.

Epstein, Larry G. and Stanley E. Zin. 1989. Substitution, Risk Aversion, and the Temporal Behavior of Consumption and Asset Returns: A Theoretical Framework. *Econometrica* 57(4), 937–969.

Fama, Eugene. 1965. The Behavior of Stock Market Prices. *Journal of Business* 38(1), 34–105.

Fama, Eugene F. and Kenneth R. French. 1996. Multifactor Explanations of Asset Pricing Anomalies. *Journal of Finance* 51(1), 55-84.

Fama, Eugene F. and Kenneth R. French. 2001. Disappearing Dividends: Changing Firm Characteristics or Lower Propensity to Pay? *Journal of Financial Economics* 60(1-2), 3-43.

Fama, Eugene F. and Kenneth R. French. 2002. The Equity Premium. *Journal of Finance* 57(2), 637–659.

Fama, Eugene F. and Kenneth R. French. 2008. Dissecting Anomalies. *Journal of Finance* 63(4), 1653-1678.

Feynman, Richard. 1955. The Uncertainty of Science. In Richard Feynman, *The Meaning of It All: Thoughts of a Citizen Scientist*. 1998. Perseus Books.

Feynman, Richard, 1963. *Lectures on Physics,* Volume 1. Addison-Wesley.

Fisher, Ronald A. 1930. *The Genetical Theory of Natural Selection.* Clarendon.

French, Kenneth. R., G. William Schwert, and Robert F. Stambaugh. 1987. Expected Stock Returns and Volatility. *Journal of Financial Economics* 19(1), 3–29.

Freund, Yoav and Robert Schapire. 1997. A Decision-Theoretic Generalization of On-Line Learning and an Application to Boosting. *Journal of Computer and System Sciences* 55(1), 119–139.

Friston, Karl. 2009. Predictive Coding Under the Free-Energy Principle. *Philosophical Transactions of the Royal Society B: Biological Sciences* 364(1521), 1211–1221.

Friston, Karl. 1010. The Free-Energy Principle: A Unified Brain Theory? *Nature Reviews Neuroscience* 11(2), 127–138.

Frongillo, Rafael M., Yiling Chen, and Ian A. Kash. 2015. Elicitation for Aggregation. *Proceedings of the 29th AAAI Conference on Artificial Intelligence*, 900-906.

Gabaix, Xavier. 2020. A Behavioral New Keynesian Model. *American Economic Review* 110(8), 2271–2327.

Gallant, A. Ronald, Peter E. Rossi. and George Tauchen. 1992. Stock Prices and Volume. *Review of Financial Studies* 5(2), 199-242.

Gandelman, Nestor and Ruben Hernandez-Murillo. 2015. Risk Aversion at the Country Level. *Federal Reserve Bank St. Louis Review* 97(1), 53-66.

Ghysels, Eric, Harvey, Andrew C., and Eric Renault. 1996. Stochastic Volatility. In G. S. Maddala and C. R. Rao, eds., *Handbook of Statistics*, Vol. 14: *Statistical Methods in Finance*, 119-191. Elsevier.

Glimcher, Paul W. 2010. *Foundations of Neuroeconomic Analysis*. Oxford University Press.

Gneiting, Tilmann and Adrian E. Rafferty. 2007. Strictly Proper Scoring Rules, Predication, and Estimation. *Journal of the American Statistical Association* 102(477), 359-378.

Goldberg, Robert S. 2015. A Methodology for Computing and Comparing Implied Equity and Corporate-Debt Sharpe Ratios. *Review of Quantitative Finance and Accounting* 44, 733-754.

Goodhart, Charles A.E. and Maureen O'Hara. 1997. High Frequency Data in Financial Markets: Issues and Applications. *Journal of Empirical Finance* 4, 73-114.

Gopinath, Gita, Josefin Meyer, Carmen Reinhart, and Christoph Trebesch. 2025. Sovereign vs Corporate Debt and Default: More Similar Than You Think. *Journal of International Economics* 144, 104082.

Green, Edward J. and Kent Osband. 1991. A Revealed Preference Theory for Expected Utility. *Review of Economic Studies* 58(4), 677-696.

Green, Jerry. 1987. 'Making Book against Oneself', the Independence Axiom, and Expected Utility Theory. *Quarterly Journal of Economics* 102(4), 785–796.

Grossman, Sanford J. and Joseph E. Stiglitz. 1980. On the Impossibility of Informationally Efficient Markets. *American Economic Review* 70(3), 393-408.

Hajdini, Ina. 2023. Mis-specified Forecasts and Myopia in an Estimated New Keynesian Model. Working Paper No. 22-03R. Federal Reserve Bank of Cleveland.

Halek, Martin and Joseph G. Eisenhauer. 2001. Demography of Risk Aversion. *Journal of Risk and Insurance* 68(1), 1–24.

Hansen, Lars P. 2007. Beliefs, Doubts and Learning: Valuing Macroeconomic Risk. *American Economic Review* 97(2), 1–30.

Hansen, Lars P. 2010. Calibration, Empirical Evidence, and Stochastic Equilibrium Models. Presentation for Institute for New Economic Thinking, April.

Hansen, Lars Peter and Kenneth J Singleton. 1983. Stochastic Consumption, Risk Aversion and the Temporal Behavior of Asset Returns. *Journal of Political Economy* 91(2), 249-65.

Harrison, J. Michael and David M. Kreps. 1978. Speculative Investor Behavior in a Stock Market with Heterogeneous Expectations. *Quarterly Journal of Economics* 92(2), 323-336.

Hasanhodzic, Jasmina and Andrew W. Lo. 2019. On Black's Leverage Effect in Firms with No Leverage. *Journal of Portfolio Management* 46(1), 106-122.

Havranek, Tomas. 2015. Measuring Intertemporal Substitution: The Importance of Method Choices and Selective Reporting. *Journal of the European Economic Association* 13(6), 1180-1204.

Havranek, Tomas, Roman Horvath, Zuzaba Irsova, and Marek Rusnak. 2015. Cross-Country Heterogeneity in Intertemporal Substitution. *Journal of International Economics* 96(1), 100-118.

Hawkins, Jeff. 2021. *A Thousand Brains: A New Theory Of intelligence*. Basic Books.

Hayek, Friedrich A. 1945. The Use of Knowledge in Society. *American Economic Review* 35(4), 519-530.

Herfindahl, Orris C. 1950. *Concentration in the Steel Industry*. Dissertation: Columbia University.

Heston, Steven L. 1993. Closed-Form Solution for Options with Stochastic Volatility with Applications to Bond and Currency Options. *Review of Financial Studies* 6(2), 327-343.

Hirschman, Albert O. 1964. The Paternity of an Index. *American Economic Review* 54(5), 761-762.

Hirshleifer, David. Investor Psychology and Asset Pricing. 2001. *Journal of Finance* 56(4), 1533-1597.

Hoeting, Jennifer A., David Madigan, Adrian E. Raftery, and Chris T. Volinsky. 1999. Bayesian Model Averaging: A Tutorial. *Statistical Science* 14(4), 382-417.

Hoffman, Philip T., Gilles Postel-Vinay, and Jean-Laurent Rosenthal. 2007. *Surviving Large Losses: Financial Crises, the Middle Class and the Development of Capital Markets*. Harvard University Press.

Horpestad, Jone B., Štefan Lyócsa, Peter Molnár, and Torbjørn B. Olsen. 2019. Asymmetric Volatility in Equity Markets Around the World. *North American Journal of Economics and Finance* 48, 540-554.

Huang, Jing-Zhi and Ming Huang. 2012. How Much of the Corporate-Treasury Yield Spread Is Due to Credit Risk? *Review of Asset Pricing Studies* 2(2), 153–202.

ICE BofA Indices. 2024a. Subsets of US Corporate Master Index sorted by credit grade or duration. Retrieved from fred.stlouisfed.org/series.

BofA Indices. 2024b. Subsets of non-US corporate credit spreads. Retrieved from fred.stlouisfed.org/series.

Ilmanen, Antti. 2011. *Expected Returns: An Investor's Guide to Harvesting Market Rewards*. Wiley.

Jagannathan, Murali, Clifford P. Stephens, and Michael S. Weisbach. 2000. Financial Flexibility and the Choice between Dividends and Stock Repurchases. *Journal of Financial Economics* 57(3), 355-384.

Jarrow, Robert, Donald R. van Decenter, and Xiaoming Wang. 2003. A Robust Test of Merton's Structural Model for Credit Risk. *Journal of Risk* 6(1), 39-58.

Jegadeesh, Narasimhan and Sheridan Titman. 1993. Returns to Buying Winners and Selling Losers: Implications for Stock Market Efficiency. *Journal of Finance* 48(1), 65-91.

Jegadeesh, Narasimhan and Sheridan Titman. 2001. Profitability of Momentum Strategies: An Evaluation of Alternative Explanations. *Journal of Finance* 56(2), 699-720.

Julliard, Christian and Anisha Ghosh. 2012. Can Rare Events Explain the Equity Premium Puzzle? *Review of Financial Studies* 25(10), 3037-3076.

Jung, Hae Won, Jin Yu, and Qi Zeng. 2020. Leverage Dynamics under Segmented Equity and Debt Markets. Available at ssrn.com 352879.

Kadane, Joseph B. 2011. Partial-Kelly Strategies and Expected Utility: Small-Edge Asymptotics. *Decision Analysis* 8(1), 4-9.

Kalman, Rudolf E. 1960. A New Approach to Linear Filtering and Prediction Problems. *Journal of Basic Engineering* 82(1), 35–45.

Karni, Edi. 2009. A Mechanism for Eliciting Probabilities. *Econometrica* 77(2), 603-606.

Karpoff, Jonathan M. 1987. The Relation Between Price Changes and Trading Volume: A Survey. *Journal of Financial and Quantitative Analysis* 22(1), 109-126.

Kelly, John L. 1956. A New Interpretation of Information Rate. *Bell System Technical Journal* 35, 917–926.

Kendall, Alex and Yarin Gal. 2017. What Uncertainties Do We Need in Bayesian Deep Learning for Computer Vision? In *Advances in Neural Information Processing Systems*, vol. 30.

Keynes, John M. 1921. *A Treatise on Probability Theory*. MacMillan.

Keynes, John M. 1936. *The General Theory of Employment, Interest and Money*. MacMillan.

Keynes, John M. 1937. The General Theory of Employment. *Quarterly Journal of Economics* 51(2), 209–223.

Kimura, Motoo. 1983. *The Neutral Theory of Molecular Evolution.* Cambridge University Press.

King, Mervyn. 2016. *The End of Alchemy: Money, Banking, and the Future of the Global Economy*. W.W. Norton.

Kingma, Diederik P. and Jimmy Lei Ba. 2015. Adam: A Method for Stochastic Optimization. In 3rd International Conference on Learning Representations (ICLR).

Knight, Frank H. 1921. *Risk, Uncertainty and Profit.* Hart, Schaffner & Marx.

Knill, David and Alexandre Pouget. 2004. The Bayesian Brain: The Role of Uncertainty in Neural Coding and Computation. *Trends in Neurosciences* 27(12), 712–719.

Koijen, Ralph SJ and Motohiro Yogo. 2023. Understanding the Ownership Structure of Corporate Bonds. *American Economic Review: Insights* 5(1), 73-91.

Kotlikoff, Laurence J. 2010. *Jimmy Stewart Is Dead: Ending the World's Ongoing Financial Plague with Limited Purpose Banking*. Wiley.

Kouba, Yosra and Skander Slim. 2019. The Relationship between Trading Activity and Stock Market Volatility: Does the Volume Threshold Matter? *Economic Modeling* 82, 168-184.

Kraemer, Nick W., Evan M. Gunter, Zev R. Gurwitz, Nivritti M. Richhariya, and Lyndon Fernandes. 2024. *Default, Transition, and Recovery: 2023 Global Corporate Default and Rating Transition Study*. S&P Global.

Kurz, Mordecai. 1994a. On Rational Belief Equilibria. *Economic Theory* 4, 859–876.

Kurz, Mordecai. 1994b. On the Structure and Diversity of Rational Beliefs. *Economic Theory* 4, 877–900.

Kurz, Mordecai. 1997. Asset Prices with Rational Beliefs. In M. Kurz, ed., *Endogenous Economic Fluctuations: Studies in the Theory of Rational Belief.* Springer-Verlag, 211–250.

Kurz, Mordecai. 2009. Rational Diverse Beliefs and Economic Volatility. In Thorsten Hens and Klaus Shenk-Hoppé, eds., *Handbook on Financial Markets: Dynamics and Evolution*. 439-506. North Holland.

Lamb, Horace. 1932. Address to the British Association for Advancement of Science. Cited in Michael Tabor, *Chaos and Integrability in Nonlinear Dynamics: An Introduction*, 187. 1989. Wiley.

Layard, Richard, Guy Mayraz, and Stephen John Nickel. 2008. The Marginal Utility of Income. *Journal of Public Economics* 92(8-9), 1846-57.

Leary, Mark T. and Roni Michaely. 2011. Determinants of Dividend Smoothing: Empirical Evidence, *Review of Financial Studies* 24(10), 3197–3249.

Li, Gang, Zhangjun Liu, Jingwen Zhang, Huiming Han, and Zhangkang Shu. 2024. Bayesian Model Averaging by Combining Deep Learning Models to Improve Lake Water Level Prediction. *Science of the Total Environment* 906, 167718.

Lim, Bryan, Jiaguo Wang and Yaqiong Yao. 2018. Time-Series Momentum in Nearly 100 Years of Stock Returns. *Journal of Banking and Finance* 97, 283-296.

Lintner, John. 1956. Distribution of Incomes of Corporations among Dividends, Retained Earnings, and Taxes. *American Economic Review* 46(2), 97-113.

Liptser, Robert S. and Albert N. Shirayev. 1977. *Statistics of Random Processes*, Volumes 1 and 2. New York: Springer-Verlag. Translation of *Statistika Sluchainykh Protsessov*. 1974. Nauka.

Liu, Yang, Claire Li, Thorsten Nestmann, and Elena Duggar. 2024. *Sovereign Default and Recovery Rates, 1983-2023*. Moody's Investor Service, Data Report, April.

Liu, Sheen, Jian Shi, Junbo Wang, and Chunchi Wu. 2007. How Much of the Corporate Bond Spread is Due to Personal Taxes? *Journal of Financial Economics* 85(3), 599-636.

Liu, Sheen and Chunchi Wu. 2004. Taxes, Default Risk, and Credit Spreads. *Journal of Fixed Income* 14(2), 71-85.

Longtermtrends. 2024. Bond Yield Credit Spreads. Draws on data from www.capitalmarketsdata.com and other sources referenced at www.longtermtrends.net/bond-yield-credit-spreads/.

Maheu, John M. and Thomas H. McCurdy. 2000. Identifying Bull and Bear Markets in Stock Returns. *Journal of Business & Economic Statistics* 18(1), 100-112.

Mandelbrot, Benoit B. 1963. The Variation of Certain Speculative Prices. *Journal of Business* 36(4), 394–419.

Mandelbrot, Benoit B. 1997. *Fractals and Scaling in Finance: Discontinuity, Concentration, Risk*. Springer.

Mandelbrot, Benoit B. and Richard L. Hudson. 2007. *The (Mis)Behavior of Markets: A Fractal View of Financial Turbulence*. Basic Books.

Marcinkiewicz, Józef. 1938. Sur une proprieté de la loi de Gauss. *Mathematische Zeitschrift* 44, 612–618.

Marsh, Terry A. and Robert C. Merton. 1987. Dividend Behavior for the Aggregate Stock Market. *Journal of Business* 60(1), 1-40.

Martin, Ian W. R. 2013. Consumption-Based Asset Pricing with Higher Cumulants. *Review of Economic Studies* 80(2), 745–773.

McComb, William D. 1990. *The Physics of Fluid Turbulence*. Clarendon.

McMahan, Brendan, Eider Moore, Daniel Ramage, Seth Hampson, and Blaise Agüera y Arcas. 2017. Communication-Efficient Learning of

Deep Networks from Decentralized Data. In Aarti Singh and Xiaojin Zhu, editors, Proceedings of the 20th International Conference on Artificial Intelligence and Statistics (AISTATS).

Mehra, Rajnish and Edward C. Prescott. 1985. The Equity Premium: A Puzzle. *Journal of Monetary Economics* 15(2), 145–161.

Mehra, Rajnish and Edward C. Prescott. 2003. The Equity Premium in Retrospect. Chapter 14 in *Financial Markets and Asset Pricing*. Handbook of the Economics of Finance. Vol. 1, 889–938. Elsevier.

Merton, Robert C., 1980, On Estimating the Expected Return on the Market. *Journal* of *Financial Economics* 8, 323-361.

Merton, Robert C. 1974. On the Pricing of Corporate Debt: The Risk Structure of Interest Rates. *Journal of Finance* 29(2), 449-470.

Meyer, Josefin, Carmen M. Reinhart, and Christoph Trebesch. 2022. Sovereign Bonds since Waterloo. Policy Research Working Paper 9906. World Bank Group.

Moessner, Richhild and Előd Takáts. 2020. How Well-Anchored are Long-Term Inflation Expectations? *BIS Working Papers Series*. Bank for International Settlements.

Moskowitz, Tobias J., Yao Hua Ooi, and Lasse Heje Pedersen. 2012. Time Series Momentum. *Journal of Financial Economics* 104(2), 228-250.

Nakamura, Emi, Jón Steinsson, Robert Barro, and José Ursúa. 2013. Crises and Recoveries in an Empirical Model of Consumption Disasters. *American Economic Journal: Macroeconomics* 5(3), 35–74.

Navier, Claude-Louis. 1822. Mémoire sur les lois du mouvement des fluides. *Mémoires de l'Académie des Sciences de l'Institut de France* 6, 389–440.

Neher, Richard A. and Boris I. Shraiman. 2011. Statistical Genetics and Evolution of Quantitative Traits. *Reviews of Modern Physics* 83(4), 1283–1300.

Longstaff, Francis A., Sanjay Mithal, and Eric Neis. 2005. Corporate Yield Spreads: Default Risk or Liquidity? New Evidence from the Credit Default Swap Market. *Journal of Finance* 60(5), 2213–2253.

Nelson, Daniel B. 1991. Conditional Heteroskedasticity in Asset Returns: A New Approach. *Econometrica* 59(2), 347-370.

Nozawa, Yoshio. 2017. What Drives the Cross-Section of Credit Spreads? A Variance Decomposition Approach. *Journal of Finance* 72(5), 2045–2072.

Ohta, Tomoko. 1992. The Nearly Neutral Theory of Molecular Evolution. *Annual Review of Ecology and Systematics* 23, 263–286.

Osband, Ian, Zheng Wen, Seyed Mohammad Asghari, Vikranth Dwaracherla, Morteza Ibrahimi, Xiuyuan Lu, and Benjamin Van Roy. 2021. Epistemic Neural Networks. arXiv:2107.08924.

Osband, Ian, Seyed Mohammad Asghari, Benjamin Van Roy, Nat McAleese, John Aslanides, and Geoffrey Irving. 2022. Fine-Tuning Language Models via Epistemic Neural Networks. arXiv:2211.01568. November.

Osband, Kent H. 1985. *Providing Incentives for Better Cost Forecasting.* Ph.D. Thesis, University of California, Berkeley.

Osband, Kent. 2001. *Iceberg Risk: An Adventure in Portfolio Theory.* Texere.

Osband, Kent. 2011. *Pandora's Risk: Uncertainty at the Core of Finance.* Columbia University Press.

Osband, Kent. 2020. *Rational Myopia: How Capital Markets Learn.* RiskTick.

Osband, Kent. 2023. Revealed Preference for Expected Utility Revisited. Available at ssrn.com 4313319.

Osband, Kent, Valerio Filoso and Salvatore Capasso. 2024. The Limits of Limitless Debt. *Journal of Macroeconomics* 79, 103567.

Ou, Sharon, Yang Liu, Kuman Kanthan, Colin Ellis, Atsi Sheth, Alastair Wilson, and William Coley. 2024. Annual Default Study: Corporate Default Rate to Moderate in 2024 but Remain near its Long-Term Average. Moody's Investor Service. 26 February.

Outreville, Jean-François. 2015. The Relationship between Relative Risk Aversion and the Level of Education: A Survey and Implications for the Demand for Life Insurance. *Journal of Economic Surveys* 29(1), 97-111.

Pastor, Lubos and Pietro Veronesi. 2009. Learning in Financial Markets. *Annual Review of Financial Economics* 1, 361–381.

Popper, Ben. 2017. Poker-Playing AI is Getting Smarter and the Humans are Getting Tired. *The Verge*, January. www.theverge.com/2017/1/25/14358246/ai-poker-tournament-cmu-libratus-vs-human-losing.

Poundstone, William. 2005. *Fortune's Formula: The Untold Story of the Scientific Betting System that Beat the Casinos and Wall Street.* Hill and Wang.

Ramsey, Frank P. 1926. Truth and Probability. In F.P. Ramsey, *The Foundations of Mathematics and Other Logical Essays*, 1931. Kegan Paul.

Read, Carveth. 1898. *Logic: Deductive and Inductive*. Simpkin, Marshall, Hamilton, Kent & Co.

Reinhart, Carmen M. and Kenneth S. Rogoff. 2009. *This Time Is Different: Eight Centuries of Financial Folly.* Princeton University Press.

Reynolds, Osborne. 1883. An Experimental Investigation of the Circumstances Which Determine Whether the Motion of Water Shall Be Direct or Sinuous, and of the Law of Resistance in Parallel Channels. *Philosophical Transactions of the Royal Society of London* 174, 935–982.

Rietz, Thomas A. 1988. The Equity Risk Premium: A Solution. *Journal of Monetary Economics* 22(1), 117–131.

Rubin, Larry, Nicholas Ranson, and Xiaokai Shi. 2009. *Analysis of Methods for Determining Margins for Uncertainty under a Principle-Based Framework for Life Insurance and Annuities.* Society of Actuaries.

Samuelson, Paul A. 1965. Proof that Properly Anticipated Prices Fluctuate Randomly. *Industrial Management Review* 6(2), 41–49.

S&P Global Ratings. 2023. Credit Trends: Global State of Play: Debt Growth Diverging by Credit Quality. www.spglobal.com/ratings/en/research/articles/230906/.

Sargent, Thomas J. 2008. Evolution and Intelligent Design. *American Economic Review* 98(1), 5–37.

Savage, Leonard J. 1954. *The Foundations of Statistics*. Wiley.

Sharpe, William F. 1966. Mutual Fund Performance. *Journal of Business* 39(1), 119-138.

Shiller, Robert J. 1981. Do Stock Prices Move Too Much to be Justified by Subsequent Changes in Dividends? *American Economic Review* 71(2), 421-436.

Shiller, Robert J. 1989. *Market Volatility.* MIT Press.

Shiller, Robert J. 2000. *Irrational Exuberance.* Doubleday.

Shiller, Robert J. 2025. U.S. Stock Markets 1871-Present and CAPE Ratio. Retrieved from iedata.xls at shillerdata.com.

Siegel, Jeremy J. 2022. *Stocks for the Long Run: The Definitive Guide to Financial Market Returns and Long-Term Investment Strategies, Sixth Edition.* McGraw-Hill.

Stiglitz, Joseph E. 2010. An Agenda for Reforming Economic Theory. *Presentation for the Institute for New Economic Thinking*, April.

Stokes, George G. 1851. On the Effect of Internal Friction of Fluids on the Motion of Pendulums. *Transactions of the Cambridge Philosophical Society* 9, 8–106.

Szpiro, George G. 1986. Measuring Risk Aversion: An Alternative Approach. *Review of Economics and Statistics* 68(1), 156-159.

Szpiro, George G. and Jean-François Outreville. 1988. Relative Risk Aversion around the World: Further Results. *Journal of Banking and Finance* 6, Supplement 1, 127-28.

Taleb, Nassim N. 2007. *The Black Swan: Impact of the Highly Improbable.* Random House.

Taleb, Nassim N. 2012. *Antifragile: Things That Gain from Disorder*. Random House.

Talpsepp, Tõnn and Marc Oliver Rieger. 2010. Explaining Asymmetric Volatility Around the World. *Journal of Empirical Finance* 17(5), 938-956.

Taylor, Stephen J. 1986. *Modeling Financial Time Series*. Wiley.

Taylor, Stephen J. 2005. *Asset Price Dynamics, Volatility, and Prediction.* Princeton University Press.

Tetlock, Philip E. and Dan Gardner. 2015. *Superforecasting: The Art and Science of Prediction*. Crown.

Thorp, Edward O. 1960. Fortune's Formula: The Game of Blackjack. *Notices of the American Mathematical Society* 7(7), 935–936.

Thorp, Edward O. 2000. The Kelly Criterion in Blackjack, Sports Betting, and the Stock Market. In Olaf Vancura, Judy A. Cornelius, and William R. Eadington, eds., *Finding the Edge: Mathematical and Quantitative Analysis of Gambling*, 163-214. Institute for the Study of Gambling and Commercial Gaming, University of Nevada.

Tsai, Jerry and Jessica A. Wachter. 2015. Disaster Risk and Its Implications for Asset Pricing. *Annual Review of Financial Economics* 7(1), 219–252.

Ülkümen, Gülden, Craig R. Fox, and Betram F. Malle. 2016. Two Dimensions of Subjective Uncertainty: Clues from Natural Language. *Journal of Experimental Psychology: General* 145(10), 1280–1297.

U.S. Bureau of Economic Analysis. 2024. Corporate Profits After Tax with Inventory Valuation Adjustment (CPATAX), and Gross Domestic Product (GDP). Retrieved from fred.stlouisfed.org/series.

Wachter, Jessica A. 2013. Can Time-Varying Risk of Rare Disasters Explain Aggregate Stock Market Volatility? *Journal of Finance* 68(3), 987–1035.

Weitzman, Martin L. 2007. Subjective Expectations and Asset-Return Puzzles. *American Economic Review* 97(4), 1102-1130.

Wolpert, David H. 1992. Stacked Generalization. *Neural Networks* 5(2), 241-259.

Wonham, Walter M. 1964. Some Applications of Stochastic Differential Equations to Optimal Nonlinear Filtering. *SIAM Journal on Control* 2(3), 347–369.

Yao, Yuling, Gregor Pirš, Aki Vehtari, and Andrew Gelman. 2022. Bayesian Hierarchical Stacking: Some Models Are (Somewhere) Useful. *Bayesian Analysis* 17(4), 1043-1071.

Yared, Pierre. 2019. Rising Government Debt: Causes and Solutions for a Decades-Old Trend. *Journal of Economic Perspectives* 33(2), 115–40.

Index

G

H

I

J

K

L

M

N

O

Made in the USA
Monee, IL
02 November 2026

67383853R00103